WHERE THE CHILDREN TAKE US

WHERE
THE
CHILDREN
TAKE US

———

How One Family
Achieved the Unimaginable

———

ZAIN E. ASHER

AMISTAD
— 35 —

An Imprint of HarperCollins*Publishers*

Page 76, "I Shall Be Released," words and music by Bob Dylan, © Universal
Music Publishing Group, used by permission, all rights reserved.

HarperCollins books may be purchased for educational, business,
or sales promotional use. For information, please email the
Special Markets Department at SPsales@harpercollins.com.

FIRST EDITION

Library of Congress Cataloging-in-Publication Data has been applied for.

ISBN 978-0-06-304883-6

22 23 24 25 26   FRI   10 9 8 7 6 5 4 3 2 1

For Specky and Tina

Action is eloquence.

— WILLIAM SHAKESPEARE

Prologue

There is tragedy in my story, but my story is not a tragedy. It is a story of grit, grace, and perhaps above all, a story of extraordinary triumph that I want to share with the world.

To be fair, it is my mother's story more than mine. That she even lived long enough to become a mother is no small miracle.

Up against soul-crushing challenges detailed in the pages ahead, Obiajulu Justina Ejiofor raised four children who shattered every expectation. Her unique parenting style, her life-changing sacrifices, and her unrelenting discipline are the reasons my brother is today an Oscar-nominated actor; they are the reasons I am a CNN anchor with degrees from Oxford and Columbia; my sister, a medical doctor; and my eldest brother, a successful entrepreneur.

People usually underestimate my mother. By all appearances, she is an ordinary woman, small in stature and quiet. She speaks slowly, with the singsongy cadence of the tiny, rural village in Africa where she grew up. She likes to ride the bus, wears modest clothes, and is often too shy to make eye contact with strangers.

She is also someone who fought with every fiber of her being for her family. She carried us through a staggering tragedy, shielded us from the violence in our neighborhood, and devoted every spare cent to our education. She barely finished high school but taught herself Shakespeare, French, and the piano—just so she could teach us. She plastered clippings of Black success stories over our walls to remind us what we could achieve. And, after ten-hour shifts each day, hosted late-night study sessions so we'd always be one step ahead in school.

My mother would tell you she did nothing special. In a way, I suppose, she's right. She simply raised us the way she had been raised, the way her parents had been raised before her, and theirs before them. In the remote Nigerian village where she grew up, this is just what parents do.

Back home, mothers and fathers sometimes go to extreme lengths—doing things you may find wacky, weird, or even a bit frightening—to elevate the lives of their children. They will go without food, sacrifice their safety, destroy appliances, even ship their kids to other continents. (Yes, all of those things happen in this book.)

You may not know much about Nigeria. Our beaches aren't featured in travel magazines. There are no safaris on our savannas. Tourists don't swarm our monuments to pose for pictures. But there is one area where we do shine. With an almost conveyor-belt predictability, we have quietly sent armies of ambitious, talented, and disciplined children to every corner of the world.

In the United States, Nigerians make up a small portion of the Black population—less than 1 percent—but represented about a quarter of the Black students at Harvard Business School in 2013. As of 2006, Nigerians were the most educated immigrant group in the United

States—17 percent held master's degrees and 4 percent held doctorates. And by 2021, three of the top five richest Black people on earth were Nigerian.

This is a culture that has, in my mother's lifetime, faced so much—civil war, ethnic cleansing, and one of the world's worst famines—and found hope with so little.

That's because we see ourselves as overcomers; people who have not folded but grown stronger under the weight of our country's painful history. Somehow that history—as raw and as complicated as it may be—has only strengthened our resolve to fight for our children.

It is that resolve that generations of men, women, and children have drawn from. As you will see, my mother is a giant among them.

This is our story.

Chapter 1

I can't remember most of what happened that Sunday in September.
I couldn't tell you what the Gospel reading was at Mass that morning or whether Aunty Fatou came over to braid my hair in cornrows, or which Culture Club song was playing on the beat-up radio in my bedroom.

None of that really matters anyway. Everything about that Sunday was so routine, so plain, so unremarkable. Until the phone rang.

My mother had been waiting for that sound since morning, never straying too far from the living room just in case she missed it. Everything she'd done that day—frying plantains, leafing through the Argos catalogue, ironing my brothers' school shirts—was all a plot to fill time.

She kept telling us to turn down the television so nothing would drown out the sound. She was anxious, fidgety; we all were.

When the phone rang at 6:30 p.m., she finally gave herself permission to exhale.

"Arinze?"

It was supposed to be my dad. He was supposed to explain why he still wasn't home; to apologize for the eight hours of worry he'd put her through.

But the voice on the other end of the line wasn't his.

This voice was nervous; it hesitated and stuttered. It took a deep breath and mumbled two sentences that brought one chapter of our lives to a swift and sudden end and started an entirely new book.

"Your husband and your son have been involved in a car crash. One of them is dead and we don't know which one."

It's human nature to fear the worst when we don't hear from a loved one for several hours, but usually, the worst doesn't happen. Usually, everyone ends up all right.

This was not one of those times.

My father and eleven-year-old brother were four thousand miles away on a father-son road trip; long-awaited quality time together after a busy summer. My brother gazing out the car window, wide-eyed and inquisitive. My father pointing and explaining: the sprawling textile markets, the street hawkers selling okpa, the overcrowded yellow buses with conductors riding on the outside. All distant flashes of rich culture, a universe away from the corner shops, brewpubs, and lollipop men that littered our neighborhood in South London.

Somewhere along that six-hour stretch of bumpy highway between my father's home state of Enugu and the buzzing West African metropolis of Lagos, the man driving my father and brother swerved into the opposite lane to cut traffic. As their car veered around a bend, it was crushed by a speeding tractor trailer. Everyone in the car was killed instantly, apart from one person in the back seat, where my father and brother were sitting.

Our relatives in Nigeria were initially told both of them had died. Then, hours later, that one had survived. Then again, that both were killed. They were still in the middle of arguing, trying to work out the facts, when someone made that dreaded call to my mother.

I was five, my eldest brother was fourteen, and my mother was four months pregnant at the time.

She hung up the phone in stunned silence. Every expression shrouded in disbelief, every movement weighed down by numbness. She prayed there'd been a mistake; prayed that perhaps in the whirl-wind of sirens and stretchers that names and identities were mixed up; that somehow her husband and son had been spared. She thought maybe if she fell asleep, she'd wake up to the sound of my brother play-ing "Au Clair de la Lune" on his recorder or my dad tapping his feet to atilogwu music in the living room.

She glanced over at me, her little girl, playing happily with a few figurines on the living room floor. Her son Obinze was watching TV. She closed her eyes.

God, if you grant me just one miracle for the rest of my life, let it please come tonight.

My parents owned a small pharmacy in a residential part of Brix-ton, South London, opposite a community of housing projects. After Obinze and I were asleep, my mother drove there in the middle of the night, oscillating between bracing for the worst and hoping for the best. She unlocked the rolling metal shutters and raided the shelves, throw-ing dozens of items into a tote bag—bandages, gauze, antiseptics, cold compresses. Her job now was to help whoever had survived.

She returned home and took her passport out of the brown envelope in her bottom dresser drawer. She tucked it into her purse, threw some clothes into a tattered suitcase, arranged for our uncle Leo to care for us, and called a taxi. By dawn she was on a six-hour flight to Nigeria. Six lonely hours with nothing to focus on besides the pain that awaited. She stared out the window at the blanket of white clouds, drawing no comfort from the heavenly fluff. As the tears fell, she understood that far below those clouds lay an impossible reality.

Lost in her trance, she barely noticed when the plane landed with a thud on the runway. As the other passengers slowly gathered their belongings, she elbowed, squeezed, and pushed her way to the front of the plane. She normally would have apologized, at least said *excuse me*, but she kept her head forward. One of her boys needed her help. Their very survival might depend upon her. This was no time to be polite. She eventually untangled herself from the airplane's clutches and scrambled to make her connecting flight.

She arrived at her final stop in Enugu three hours later. The looming moment of truth made it hard to breathe as she navigated the rush of activity in the arrivals hall: throngs of people hanging around the baggage claim, embracing families and barking taxi drivers. Barefoot children sold groundnuts, and area boys offered to carry her bags.

She clutched her overstuffed tote and focused on her feet—*small steps forward*—to keep from collapsing when a young driver approached.

"Where you going? By yourself? You have more bags?"

She mumbled something about a hospital near the main market. They drove there in silence.

After the car accident, the passengers were all assumed dead. Their bodies were flung one by one into the back of a truck and driven to a local morgue. It was only when the driver arrived at the morgue, opened the back of the truck, and began unloading the bodies, that he noticed one of them was still breathing.

My mother didn't know any of that as the car pulled up to the hospital's main entrance. The concrete bungalow was swarming with people, as most good hospitals in Nigeria usually are. She pushed her way through the crowd outside and into the waiting area, her gut heavy with dread. She scanned the lobby: pregnant women fanning themselves in the brutal heat, patients and clerks arguing over hospital bills, the sick and wounded groaning from wooden benches while others slept on the floor.

She raced over to an unruly line at the front desk, unable to bear not knowing her fate for a second longer. After only a few moments, she flagged down a passing nurse.

She gave the nurse her name and was told to wait. After twelve hours of traveling across two continents, twelve hours caught in a whirlpool of fear, she now had to sit and wait. The next few minutes felt like decades. She sank into a chair, caressed her pregnant belly, pulled a rosary from her purse, and squeezed the beads. Then an attendant appeared and asked her to follow.

The hospital was laid out in a series of bungalows connected through a maze of outdoor walkways. My mother looked down as she walked over the chipped, concrete floors. The fluorescent lights overhead flickered and buzzed, and the hum of a power generator was the only background music.

They continued until they reached a closed door that led to Ward 7. She followed the attendant into an open room lined with more than a dozen hospital beds, all of them occupied.

She paused on the threshold to take in the scene: monitors beeping, nurses scurrying, patients shouting for attention. Relatives were sleeping on the concrete floors next to their loved ones' bedsides. She scanned the room twice, her heart pounding in her throat, as she searched for a face she knew. And then she saw him. A small boy, her small boy, lay helpless on a bed against the wall. She recognized his brown eyes peeking through the bloody bandages that cocooned his face. For the first time, it was real. Her husband was dead; her son was alive, but barely.

Delirious with anguish, she took one step toward her son and fainted.

She awoke sitting in a chair next to the bed. As her vision began to fill in, it took all her strength not to scream.

As the doctor explained her son's many injuries, her mind drifted toward the husband she'd never see again. Arinze Sylvester Ejiofor was her everything, her partner since the age of fourteen, the only boy she'd ever kissed, and the unstoppable force that held her family together. He was a larger-than-life character everywhere he went: an accomplished singer in Nigeria, a trainee doctor in Mexico, an aspiring entrepreneur in England. His energetic charm and wit generated a contrail of love and support that was almost visible behind him.

My mother tried to remember the last thing they said to each other. Something about needing more toothpaste or diapers for the pharmacy. They were always running low on diapers.

When the beeping machines finally broke her trance, the doctor still hadn't stopped talking. She had to concentrate hard to unpeel the string of technical terms he used to describe her son's condition: badly broken bones in his right arm and a serious head injury. From what the doctor was saying, it was clear her son had only just managed to survive. She leaned over her eleven-year-old boy and gently caressed his hand.

"It really hurts," my brother said, trying unsuccessfully to move his arm on his own.

She scrambled for the tote bag and began laying out Band-Aids and gauze on the bedside table, but nothing she'd brought could ease his pain, or hers.

The hospital was one of the most sophisticated in eastern Nigeria, but that gave my mother little comfort. The machines were old, the buildings poorly maintained, and there were serious concerns about hygiene and infection. Mosquitoes had free rein. The bathrooms rarely had running water. She let out a desperate sigh and reached for the bottle of water sitting on the floor.

"Drink this," she told her son. "You have to keep hydrated."

A nurse walking by asked her to keep her voice down so she wouldn't wake the other patients. Fighting back tears, she asked my brother how on earth this could have happened. It took him a few moments to respond. He was clearly still in pain. He spoke slowly, in a low voice, without any expression.

They were on the road to Lagos, had barely left Enugu, in fact, when my father pulled out a wooden comb from his back pocket to style his hair. It was the same comb he always used. After working on his short afro for a few seconds, he smiled at my brother. "Just making sure I look my best when I see your mom."

Seconds later, everything went dark.

⚘

The car didn't have any working seat belts, and even if it had, it likely wouldn't have mattered in a head-on collision with a tractor trailer. In the 1980s, Nigerian roads were mostly single-lane highways, poorly maintained and riddled with potholes. Few drivers paid attention to traffic rules. Death from road accidents was so common that some people banned family members from driving after sundown.

My brother Chiwetel lay in that hospital bed for over a week without knowing his father was gone. He noted it was strange that everyone had visited him apart from his dad, but family, friends, and medical staff were too afraid to break the news.

"He's recovering just like you. He'll stop by as soon as he's better," my mother said through her tears.

"Why can't I go and visit him?"

"The doctor says you're not well enough to leave your bed yet. You'll see your daddy when you're strong again," another uncle explained.

Each of his visitors conferred with the next about what they were telling my brother, so their stories matched.

Obinze and I arrived in Enugu later that week with no idea how bad it was. We knew there'd been a car accident, a serious one, but no one told us what had happened to Dad. We arrived at our grandmother's house believing he was going to pull through, that he'd end up all right.

When we were finally reunited with our mother, the expression on her face told us otherwise. She motioned for us to sit next to her on the faded brown couch in the living room. We could hear by her voice she'd been crying. Obinze's eyes went wide with fear.

"It's worse than we thought. The doctors tried." She paused for a few moments. "Dad didn't make it."

Obinze began to howl. I cried tears of fear and confusion. At such a young age, I didn't realize that *didn't make it* meant I'd never see my father again. I knew something bad had happened to him, something bad enough to make my mother and brother cry, but no one explained exactly what. She wrapped Obinze into her chest and squeezed my hand hard as we all wept together. The more they cried, the more I cried.

The next day my mother returned to the hospital and told her youngest son the truth. All week he'd been talking about his dad as though he were still alive.

"Can you ask Dad if he still has my comic books in his bag? Can you ask him if we're still going to Alton Towers next weekend?"

All week everyone around him pretended, too. "Yes, we'll ask him next time we see him."

When my mother finally told him, he shook his head in disbelief. "No, he can't be—no. No. No."

She reached for his hand as pools of tears began to form in her own eyes. "I wish this wasn't real, but we have to be strong. He would want us to be strong."

They sobbed and shook on that dingy hospital bed as though they were the last two people in the world.

For a time, the overpacked hospital ward became a second home of sorts for my brother. Aunties, uncles, and distant relatives from our village constantly cycled through, shedding tears at his bedside and crying for hours in the hallway with my mother.

But my brother mostly shared his days and his pain with the other patients in his ward—swapping stories, sharing food, and crying the same tears when death inevitably came. Having already lost his father, my brother watched six others die in that hospital ward that fall. And every time someone left, another patient, fighting for his or her life, appeared within hours.

While caring for her son, my mother was forced to decide how best to say goodbye to the only love she'd ever known. Her husband was just thirty-nine years old, yet his life story was already peppered with superlatives. She wanted his funeral to be as memorable as he was. She needed it to stand out and leave a mark the way he had.

A few days before the service, she asked one of her sisters to shave her head, as is the tradition for widows in Igbo culture. Thick puffs of my mother's jet-black hair fell to the ground with every swipe of the buzzing razor, her naked crown a jarring reminder of the life-shattering loss she had endured.

Later in the afternoon, more sisters came by with bags stuffed with white blouses, skirts, and dresses—another Nigerian tradition she wished she didn't have to practice. Widows in our tribe are often encouraged to wear white for a full year after the funeral to honor the life that was. That night, hairless and covered by a donated white dress two sizes too big, she looked up from her plate of jollof rice and asked her mother for help with the funeral guest list. She knew this would take the most time. Nigerian funerals are truly celebrations of life. Almost anyone who has ever met the departed is invited. For someone as sociable as my dad, the list could easily surpass one thousand.

The morning of the funeral, the hearse carrying my dad's casket led the procession past the places that meant the most to him while he was alive: his parents' tiny bungalow in Imama, the square where he danced atilogwu with friends, the field where he played soccer. Mourners lined the streets at each stop, singing, wailing, and dancing. It was the kind of display that might be reserved for a head of state in the West, but in the village, this is how it's always been.

During the service, my mother sat expressionless beside the coffin as the people who cared most about our family read speeches, sang hymns, and cried. She looked hollow, vacant, as if her soul had been scraped from her body. I played with one of my little cousins in the back, still trying to figure out why we were all there. I saw people hunched over, burying their heads in their hands, while others sobbed in front of the open casket, where Dad looked like he was taking a nap. I spent the entire service playing, relieved that my mother was far too distracted to drag me to my seat and tell me to sit still.

The parish priest was soft-spoken and tall. He gestured solemnly as he read from the book of Matthew.

"Come to me all ye who are weary and burdened and I will give you rest."

The cries of relatives drowned out much of the sermon, as did the rain banging on the zinc roof. I watched through the window as the water poured from the gutters, forming mini waterfalls that landed in big mud puddles. The heavens were crying with us.

My mother's life wasn't supposed to go like this. Her husband was on his way to setting up his own practice with steady patients, on-call hours, and a framed certificate on the wall. My parents sometimes stayed up well into the night talking about saving to rebuild this very village they grew up in, how they would bless it with running water, electricity, perhaps even a library. Imagine that: Oyofo-Oghe with its own library, they'd once said, laughing. Maybe they'd retire to Nigeria in twenty or thirty years and move into a duplex in a fancy neighborhood if they saved well and spent wisely. They were going to make a name for themselves. Their entire village would be proud.

After the Mass, a group of men carried the casket on their shoulders up a hill all the way to a burial site in front of the house my father grew up in. Our villages have no cemeteries, and the dead are typically buried outside their childhood homes. My mother, visibly pregnant, followed the procession, her eyes wilting under a fog of sorrow, until she reached a one-story home with a white door. It looked smaller than she remembered from her teenage years. Most of it was in disrepair, the mosquito net at the front door torn, the outside walls stained with mud and soot. They had once dreamed of fixing this place up, too.

After the burial, we traveled back to my mother's childhood home, a tiny two-room bungalow in Ugwuagor where her parents moved after the war. The eldest of nine, she and her siblings lit candles, prayed, and traded their fondest memories of Dad. More than once, my mother slipped away to the bathroom, where she retched and heaved as quietly as she could. The morning sickness had lingered into her second trimester, a constant reminder that a new life was growing inside her, innocent and unaware of its father's death.

A transistor radio crackled in the background as they prayed. During one evening newscast, the announcer noted that an accomplished doctor and musician was buried earlier in the day in his village. He'd been killed in a car crash but somehow his young son miraculously survived.

That was the first time my mother heard news of her nightmare spoken publicly. Her younger sister rubbed her back.

"Please check on Chiwetel at the hospital," my mother whispered. "He had surgery today."

ﾒ

Grief is a familiar, if unwelcome, visitor to many Nigerian homes.

Before she lost her husband, my mother had already survived a brutal civil war, the death of her younger brother, and a two-year famine. But that evening after my father's funeral, as she lay in a bed crowded with family, she struggled to accept the weight of her latest calamity.

She had no inkling of what the future held. She could only think of the past, her husband, and what was lost. She stared up at the concrete ceiling of her childhood home, wiped her tears, and pored over the events that led to that devastating moment.

Just a month earlier, her entire family was safe back home in London. She was in the kitchen cooking a stew one night as her husband leaned against the fridge, passionately explaining why a father-son road trip in Nigeria was so critical. He tried to keep his voice down so as not to disturb me and my brothers.

"It's just an extra week," he tried to reason.

She pushed back. "It's just not worth it."

"He needs this. He needs to understand where he's from. Do you know what names they call him at school?"

My mother knew all too well. Chiwetel, one of the few Nigerian children in his class, was being picked on for being different. She'd spoken to his teachers twice, even tried to confront one of the offending children's parents, but nothing changed.

My father suggested that if his son knew more about his heritage and culture, he'd find it easier to stand up for himself. He proposed a road trip across Nigeria, touring the same villages and towns that had produced centuries of proud Nigerians.

My parents argued back and forth into the night.

"Why travel thousands of miles for something he can learn in a textbook?"

"He'll remember this trip forever. There's no better gift."

My mother eventually gave in.

Early the next morning, my dad burst into Chiwetel's bedroom, clapping his hands in delight as he split the curtains in half. He announced they were going on the trip of their lives, quickly reeling off a list of places they could visit: the ancient walls of Kano, maybe the Olumo rock near Lagos, or the Onitsha market in the east. They were going to study Nigeria from top to bottom, left to right, explore every corner. His son was going to know who he was.

My parents were already invited to a wedding in Nigeria that summer. We decided to make a family trip out of it, and Chiwetel and Dad would peel off afterward for a week alone together. He booked the flights the following day.

We landed in Kano on a humid afternoon in August, our suitcases stuffed with gifts for everyone we knew. Some of the items were new, but many were hand-me-downs: sleeveless patterned dresses, Levi's jeans, Converse sneakers for our cousins. In the '80s, anyone who'd left Nigeria for London and actually made a decent living was treated like royalty when they came back. Soon after we arrived, relatives gathered around and bombarded us with questions about life in the West.

"How often do you have running water in England?"

"Can you show us pictures of snow?"

My dad took us to visit old friends and extended relatives later that week. A bevy of uncles and aunties we'd never met pulled us toward them, squeezed us, drew back, caressed our cheeks, and squeezed again. It was as if they'd been waiting their entire lives to meet us. They joked about my dad's high school years and, with deep earnestness, told us to always remember that this place, Nigeria, would forever be our true home.

My father spent hours talking with his friends about the days before the war, when Nigeria glistened like the jewel it was meant to be. Back then, Lagos could have been mistaken for London or New York, they insisted proudly. They swapped views on whether President Babangida would turn Nigeria into a true democracy, the Super Eagles' impressive run at the Africa Cup of Nations, and how desperately the Wawa people needed their own state. My dad, brimming with ambition even then, planned to leave his mark, perhaps by starting a school or a hospital somewhere in Enugu.

The wedding day came quickly. The five of us arrived at St. Louis Catholic Church in Bompai just as the sun had taken prime position in the cloudless sky. As the doors opened, we joined a chattering parade of color; red lace wrappers, green agbadas, blue dashiki shirts flowed slowly into the dome-shaped structure. My father stood out in his dark gray jacket, as did my mother in her elegant red chiffon dress, choices that hinted at their budding success in the West.

Once the newlyweds were announced, everyone was keen for the party afterward—no one more than my dad. He was famous on at least two continents for his shenanigans at parties. I loved to watch him dance.

We moved from the chapel to an adjacent hall. After the speeches were over, and the music grew a little louder, my father draped his jacket over the back of his chair and shimmied to the center of the dance floor. He was alone at first, but didn't care. He took off his shoes, loosened his tie, crouched down and began to bounce. Then he spun, jumped, and shook his shoulders. He was smiling. The others quickly circled him, pointing, clapping, and laughing as my father twisted, turned, and tapped his feet with increasing speed. My mother, mouth agape, shook her head in disbelief, then slowly relaxed and smiled. This was typical Arinze, after all.

The next morning was my parents' last together.

They woke up giggling like teenagers about the night before. My mother sat upright and impersonated his dance moves, the head bopping and the finger clicking; my dad erupted in belly laughter. She nudged him to keep his voice down, as the rest of the house was still sleeping. They chuckled under their breath.

At breakfast, we munched on pap and akara as my dad and Chiwetel hammered out the final details of their trip, sketching out a rough map of Nigeria and circling all the places they'd see once they boarded a bus and headed south. My mother, Obinze, and I would fly back to London that evening, leaving the two of them to begin their cross-country adventure the next day.

At dusk, we rolled our suitcases along the gravel driveway toward the humming engine of a taxi. My father loaded our belongings into the car, slammed it shut, and gathered my mother in his arms for a final embrace. She felt the steady pitter-patter of his heart.

She wiped away her tears with the sleeve of her dress. "You're sure you don't want to just come back with us?"

She didn't like the thought of leaving Dad. None of us did, but she'd already given her blessing. She climbed into the passenger seat, closed the car door, and rolled down the window. If she'd known these would be her last words to him in person, perhaps she would have said something else, something more meaningful.

"Call us when you get to Enugu," she said, squeezing his hand.

"I'll be home soon," he said, kissing her forehead.

She rolled up the window as the taxi pulled away, straining her neck to catch one last glimpse of her husband. He'd already disappeared into the house.

My dad and Chiwetel set off early the next morning.

They boarded a bus heading south to Enugu—five hundred winding miles of rugged, bumpy terrain alongside a mass of humanity. Women with babies wrapped to their chests, sleepy-eyed students returning from summer vacation, and young couples looking for a fresh start. My brother slept on my dad's shoulder until gently nudged awake as they began to pass important monuments and landmarks. They stared out the window over the hilltops at the green rural expanse flung out in front of them. Nigeria's natural beauty looked almost indulgent. My father pointed west toward the Emir of Zazzau's palace, laughed with pride about the Igbo warriors whom the British invaders found (almost) impossible to conquer, and lamented the lost kingdom of Benin, whose prized brass statues were looted by British troops before they burned the city to the ground.

When father and son arrived in Enugu, they spent a full day at Iva Valley, where more than a dozen coal miners lost their lives protesting against the British in 1949. The massacre helped spur the fight for the nation's independence. Later that week, they headed south toward the National War Museum in Umuahia, where my father recounted the heroism of ordinary men he saw with his own eyes during the Biafra war. Students, farmers, and day traders became soldiers almost overnight, fighting well-armed government soldiers with stones, machetes, and homemade guns in a fierce but short-lived battle to preserve the Republic of Biafra.

In the days that followed, he shared countless stories of their ancestors' bravery, their unwavering rebellion in the face of injustice, and the bold beauty of the wild jungles that framed the landscape. The message was unmistakable: No matter what those kids at school said, Nigeria was a land of heroes.

Back in London, we counted the days until their return. The house felt bare without my dad, his warm energy and booming laughter. We always said that a room never felt more empty than just after he'd left it.

A few hours before the accident, he called home with an update. Obinze and I ran downstairs and huddled around the receiver. He and Chiwetel were just about to make the six-hour journey to Lagos: the final leg of their weeklong road trip.

"This has been good for him. He won't forget this."

My mother reluctantly admitted her husband was right.

And just before he hung up, a quick reminder: "Don't forget to order more diapers and toothpaste for the pharmacy. We're out again."

"Already done," she said smugly. "See you tomorrow."

She never heard his voice again.

Chapter 2

My mother fell in love with the first boy she ever danced with. She was shy and unassuming; her knowledge of romance was limited to the young adult novels she searched out at the school library.

She first laid eyes on Arinze in the summer of 1965. She was fourteen and the oldest of seven by then. He was working at the most popular restaurant in the village, where he swept floors and cleaned tables with a cloth that hung from his back pocket. Arinze was the type of sixteen-year-old who spent more time talking to customers than actually cleaning.

My mother was one of those customers. She was dining that night with her cousin Evelyn, who knew Arinze well and called him over to say hello. Arinze was immediately captivated by the tall, slender girl with the bashful but inviting gaze. She was beautiful in a regal kind of way. Somewhere between the gentle curve of her eyes and the slope of her cheekbones lay a quiet self-assurance. He'd never seen features arranged so carefully.

"Have you met Obiajulu?" Evelyn asked, gesturing toward my mother.

"I don't think so." Arinze was staring.

"She's home for the holidays. Her family lives up north."

Captivated, Arinze dropped the cloth he was holding. He reached out to shake her hand.

"I'm Arinze."

She repeated it back slowly. *Ah-rin-zeh.* Saying it made her feel warm. She sized him up. He was a jovial young man, broad in both size and personality, and he had kind eyes.

Obiajulu didn't have much time for boys in those days. Just thirteen years her mother's junior, she was responsible for commanding a small army of younger brothers and sisters back home. She was far too focused on family duties to waste time with strange new boys, but her sense of responsibility softened in Arinze's presence. She felt drawn to him.

After only a few minutes of small talk, Arinze shouted to his boss that he was leaving early. He ripped off his apron, clapped the dust from his hands, and ushered the girls out onto the street before anyone could stop him.

The village of Oyofo-Oghe was an 8,000-member family tucked away in the rural expanse of western Enugu. There was no electricity or running water in those days, and Obiajulu, like the other girls, trekked more than a mile by foot a few times a week to fetch water from the Ajalli River. Like virtually all the villages that dotted the countryside beyond Nigeria's sprawling cities, every individual family was connected—if not by blood, by unwritten rules that required a deep sense of shared ownership, responsibility, and pride in the community's successes and failures.

Arinze, Obiajulu, and Evelyn wandered into the busy village square shortly before nightfall. A group of boys were singing a song that only Arinze recognized. It was the Beatles' "I Want to Hold Your Hand." The girls listened with curious skepticism as Arinze told them about the British musicians, and several others from America: the Everly

Brothers, Bob Dylan, and Sammy Davis Jr. He was showing off, and it was working.

Who is this young boy whose mind is swollen with so much knowledge? Obiajulu wondered.

When the music stopped, Obiajulu realized she'd been staring at him longer than she should have. Arinze looked down at her and smiled.

My parents would later describe their meeting as love at first sight, but neither would have admitted it back then. That first evening beneath the darkened sky was a slow dance of thoughtful conversation and cloaked teenage excitement.

They spoke hopefully of Nigeria's future, its recent independence from Britain, and Arinze's plans to travel abroad—Latin America and maybe Europe. Obiajulu joked she'd tag along. She sometimes made the daylong journey between Enugu and Jos with one of her aunties, but she'd never really contemplated a life beyond Nigeria, and certainly not a life beyond the continent.

They didn't notice when Evelyn wandered away, but finding themselves alone in the dusty village square, they joined hands and began to sway back and forth. There was no more talking.

The English translation of the Nigerian name Obiajulu is simply: *My heart is at peace.* And so it was that night, dancing under a canopy of stars as the village boys sang another tune from a distant land she didn't recognize. It dawned upon Obiajulu that she'd probably remember this night for the rest of her life. And she does.

From that moment on, they were inseparable.

~

After the funeral, we returned to London without Chiwetel.

The doctors in Nigeria warned us he was too fragile to be moved, that dragging an eleven-year-old out of the hospital so quickly after a serious car accident would be risky. My mother yelled and cried and begged, but the doctors were adamant. He needed at least another month, maybe longer, in the hospital.

Obinze and I had to get back to school, and the pharmacy in Brixton was bleeding money. She had no choice. My mother instructed her brothers to watch over him carefully and to call every few days with updates.

We stopped by the hospital to say our goodbyes, each of us leaning in slowly to embrace him. The life had begun to return to my brother's eyes, but the entire left side of his body was in agony and he still couldn't move his right arm. Thick bandages covered the ghastly wounds on his head, and a plaster cast concealed his forearm. My mother caressed his cheeks and promised she'd be back to fetch him before he knew it.

We arrived in London early on a Thursday morning. None of us had slept on the flight. It still felt as though we were living in a dream, a very bad dream. My mother was trapped in a daze as she led us through the terminal toward the exit and into the back seat of a waiting taxi.

The city was just beginning to wake as the early sun filtered in through the car window. On any other day, Obinze would be singing along to a pop song on the radio, but there was no singing that morning. The previous two weeks had aged him, his eyes wilted, his face stripped of innocence. He stared blankly out the window at the world flashing past.

As we drove home from the airport, my mother—in no mood to make small talk with the driver—glanced back constantly to make sure we were still there, still touchable. She was only just beginning to realize she could have lost two members of her family that week. Somewhere deep within the grief was a speck of relief.

Even at that early hour, traffic on the M4 moved sluggishly. The city beyond the windshield seemed desperately foreign, just as it was when she first arrived two decades earlier. She couldn't fathom this city without her husband. She had only ever seen it through his eyes, through his plans and observations. She couldn't bear to see it alone through hers.

They'd landed in London as teenagers in the early '70s, overwhelmed by the city's many pitfalls and promises. They were Black immigrants in an overwhelmingly white city that was slow to embrace multiculturalism. They'd never experienced life as a *minority* before, and they knew, just like everyone else, that their Britishness came with an asterisk, that no matter what the local politicians said to win their votes, they'd only ever *half* belong. They traded notes on how to adapt and reassured each other whenever they felt unwelcome, overwhelmed, or confused, which was often.

Through the blur of her tears, it suddenly looked as though London were underwater. She stared out the taxi window, focusing intently on the broken white lines on the asphalt below. Her head was still in Nigeria, with the son she'd left in the hospital and the husband freshly buried in his village.

She scanned the city streets. London was coming to life as we drew toward our home. Commuters flowed in and out of underground stations, walking with purpose and self-importance as they headed to their classrooms, boardrooms, and business meetings. She used to be like them. She used to have direction. She used to walk with purpose. All that seemed so trivial now.

We drove across Battersea Bridge, where she remembered the walks she and my dad took in the nearby park where the water swooshed against the banks. It was up there where Arinze had first shared stories about major British landmarks: that Big Ben was likely named after a famous boxer, and that even now, there were dozens of real families living in the Tower of London. He loved to impress his young wife with his knowledge of British history and culture.

We continued farther south through Clapham, passing the large semidetached homes near the Common they used to gawk at on their way home from work. "Stick with me and we'll be living in one of those soon enough," he'd joked.

Eventually, we passed the registry office where she and my father signed their marriage into law. The wedding was so clinical and bureaucratic—not one of those boisterous, dance-till-you-sweat African weddings with hundreds of guests, kola nut rituals, and traditional dancers. She'd grown up attending weddings like those, weddings that literally drew the entire village and lasted two days. Instead, she got a small registry hall surrounded by a few familiar faces on a freezing Saturday morning in January. Now, suddenly a widow, she squeezed that memory with all her heart to keep it from slipping away like he had.

None of us wanted to get out of the taxi when we pulled up to our small terraced home on Tulsemere Road. Obinze, now the man of the house at fourteen, climbed out first as the taxi driver opened the trunk. He removed four bags, including the suitcase my father carried with him on that road trip to Lagos, and dragged them toward the house.

My mother ushered us over the threshold, gently closed the front door, and turned on the lights. She scanned the family photos on the mantelpiece, the notepads filled with lyrics to Dad's songs, his empty chair at the head of the table. His coat was still draped over the banister, his spare keys on the side table in the living room. He once lived here, he once shuffled through these halls, this space was once his.

This house, just weeks earlier bursting with the energy of three boisterous children and my dad's dynamic presence, was cold and silent. Weighed by a feeling of numbness, it dawned on my mother that we were now a family of four—three, actually, until Chiwetel returned.

She asked Obinze to carry our father's bag to their bedroom. They were both aware of the emotional weight of her request, but no other words passed between them. He nodded, dragged the bag upstairs, and retreated to his room for the rest of the day.

I ran inside and played alone on the living room floor with a train set my dad had given me a few weeks earlier. I asked my mother when he'd be home to play with me. She closed her eyes and shook her head before climbing upstairs to a room that was suddenly hers alone, despite the ever-present reminders of the man she'd just lost. A tornado of panic swelled inside her as she remembered the last time she was there, packing her bags in the dark, and the stuff of nightmares that came afterward.

She drew the curtains closed and crawled into bed fully dressed. She thought about the baby growing inside her—what she'd say when he or she was old enough to ask about their father. Three weeks had passed since the accident, and my mother had missed multiple doctor's appointments. She knew nothing of the health of her unborn child, whether it was coping with her irregular eating and intense anxiety. She reached into her pocket and yanked out a sheet of pills a doctor in Enugu had prescribed for the stress. She shook two capsules into her palm, tossed them to the back of her throat, closed her eyes, and swallowed. She pulled the covers over her face and tried to fall asleep.

She thought back to that morning before the trip to Nigeria, how Arinze had nudged her awake an hour early with a burst of fresh energy in his voice. The wait was finally over. "This time tomorrow, we'll be home."

Arinze had a habit of making everything seem much more exciting than it actually was. He lived for adventure. And Nigeria, although dangerously disorganized, was always an adventure. There was always some commotion to gawk at, some bizarre scam not to fall for, and if you didn't come back with at least two or three funny stories to share, you clearly hadn't been paying attention. But more than anything, Nigeria made him feel free. It had that effect on people. If he wanted to collapse in laughter in the middle of the market or entertain a wedding crowd with his atilogwu dancing, he could—a far cry from how he was expected to behave as a trainee doctor in London.

He loved spotting a group of children playing soccer in the village, and joining in to dribble the ball and score a goal, or tapping people on the shoulder he hadn't seen in years, and watching them scream with joy once they turned around and saw him. My mother smiled as she remembered Arinze grinning from ear to ear the day of their trip, leaping out of bed like a child on Christmas morning. She fell asleep cradling the memory.

She awoke at noon the next day. It took all her energy to haul herself out of bed. My mother, who normally dressed meticulously, was still wearing the same white dress and baggy sweater she'd worn on the flight home from Nigeria. She looked disheveled, drunk with grief, and there were bulging pouches under her eyes. She squeezed the banister as she padded downstairs, her eyes cast down at her feet. She spotted Obinze in the hallway, handed him a few pounds, and mumbled something about groceries. Bread, rice, beans, sardines—anything that didn't require thought. She trudged back up the stairs, closing the bedroom door behind her. That was as much human interaction as she could take that day.

Those early days were the darkest.

We settled into a frightening new routine. My mother would come downstairs each morning to make sure we had food to eat, then she'd disappear until the following day.

I don't know how many days she lay in bed drifting in and out of that nightmarish slumber. Some days she'd slip out of her bedroom at 2 a.m., others at 5 p.m. In those first weeks back, days lost their meaning—time mattered only to help count the days since her husband's death, September 3, 1988.

ー

Obiajulu and Arinze were married at an uneventful registry ceremony in Brixton, in the winter of 1974. They'd already promised each other eternal love and loyalty back in the village three years earlier, but in the Western world you needed a certificate to prove *everything.* On the day of the wedding, Arinze wore his best native suit and matching hat and was up early polishing his shoes by the door. Obiajulu squeezed herself into a matching traditional patterned blouse and skirt handmade for the occasion. The weather was so cold she had to wear thick tights underneath and one of my dad's blazers, which ruined the entire look, but none of that bothered her. She and Arinze were destined for an eternity together. The outfits did not matter.

As soon as everyone was seated, the officiant quickly got down to business: "Do you, Ay-rin, take Obi-joo to be your lawfully wedded wife?" Obiajulu and Arinze couldn't help but laugh. They were used to British people twisting themselves into knots trying to pronounce their names, but they were far too focused on the matter at hand to care. Arinze, still smiling to himself, gently pulled Obiajulu closer and nodded, "I do." And of course he did. They'd been waiting nine long years for this day. Though separated at times by oceans, forests, and war, they always knew that whatever that special feeling was, that unyielding covenant that connected them, it was far too powerful and too persistent to let them be apart for long.

ー

After a week back, my mother hauled herself out of bed and moved toward the stuffed brown suitcase in the corner that had somehow survived the accident that claimed her husband.

She reminded herself to breathe. She opened the bag and began laying out several items neatly on the floor, wrestling with what to keep and what to give away. His clothes still smelled of Enugu: the wet heat, the clay roads, Omo soap. She pulled out a pair of grass-stained jeans, a few crumpled shirts, and a leather-bound address book he'd started five years ago. She flipped through the small pages, realizing there were people she still needed to notify. A former professor he stayed in touch with, a Nigerian singer he befriended during his own music career, classmates from the University of London.

She shuffled over to the wardrobe and wheeled out a portable filing cabinet containing their most important documents: college acceptance letters, diplomas and degrees he barely got to use, and every single love letter they'd written each other over the past two decades. The carefully handwritten pages detailed the origins of their romance, that first dance in the village square, and their struggle to survive the civil war that erupted two years later.

Though they were apart for several years during that war, they remained committed to a future together, eventually moving to Europe as soon as it was safe to do so. First to Paris, where Arinze finished his studies, then London, where Obiajulu enrolled in pharmacy school. They'd walk home together after her classes at King's College, laughing about life in the West and all they wanted to experience.

Even back then, Arinze had a clear idea of where he was going. He only wanted to return to Nigeria after England had fulfilled all of its promises, after he had a *Dr.* in front of his name, after he'd made the people who raised him proud. He wanted to open another pharmacy, maybe move to America one day, retire to Lagos, and tour the world in his spare time. While many of his friends ran back to Nigeria complaining of the cold winters and brutal recessions, Arinze refused to leave empty-handed. He was stubbornly optimistic. He carried himself with such a shiny, metallic confidence that if all his thoughts about who he was and where he was going were printed on his forehead, he'd have nothing to be ashamed of.

During those hours alone in her bedroom going through his things, she'd sometimes convince herself he was about to come bursting through the door. More than once, she heard the sound of his car pulling up outside, his keys jingling in his pocket, and his footsteps dancing up the stairs. She imagined jumping into his arms and spending the night laughing and sifting through photos of his adventures. She stared at the closed bedroom door waiting for it to open.

She carefully tucked the letters back into the filing cabinet, folded his clothes, and placed them back into their wardrobe. She was going to keep everything.

🖎

My brother and I survived those initial weeks with the help of extended relatives. On weekends, our living room was filled with a parade of weeping aunties and sympathetic uncles giving us food, flowers, and hugs. They promised we would be okay. But my mother was still locked away upstairs, leaving Obinze to greet the mourners alone. He offered them water, accepted their condolences, and made sure they were comfortable. Everyone wanted to know how we were doing, whether we were coping. They could see the dishes piled up in the sink, the days-old half-eaten sandwiches on the table, and our mother's glaring absence.

We later learned that our extended relatives held a meeting to share their concerns about our mother. Many of them had known her since childhood; Obiajulu was gracious and dutiful but she wasn't strong enough to withstand a tragedy like this. She needed help. They sent word back to Nigeria and our grandmother arrived in London shortly afterward.

Our grandma, known to us as Mama Nnukwu, an Igbo term that literally means "Big Mama," was tough and nurturing in a biblical kind of way. With barely any formal education, she'd grown up in a village where she hunted her own food and started a family at thirteen years old. She, too, had seen firsthand the pain that life could serve when one of her sons, a boy of just eight, died from the shortage of food and medicine that devastated our people during Nigeria's civil war. And like most victims of conflict, she barely had time to grieve. The next day, as enemy soldiers approached and bombs fell, she gathered her remaining children and fled into the bush.

Hardened by such suffering, Mama Nnukwu had little tolerance for fuzzy emotion. She had come to London for a single purpose: to help her daughter stand on her own again, and there wasn't a minute to waste.

After several days of listening to the screams and wails on the other side of my mother's locked bedroom door, my grandmother forced it open and ordered her out of bed.

"Hiding from the world won't bring him back. Look at your children—they need you."

She seemed to listen, but little changed at first. It took almost a week before she joined us for dinner. She waded downstairs into the dining room one evening, wearing one of Dad's old white T-shirts. She slid into his old chair and slurped her soup in silence, her shoulders hunched over. We were afraid to make eye contact.

My grandmother soon persuaded her to venture outside for the first time in weeks. She wrapped her heavy winter coat over her pajamas one evening, twisted a scarf around her neck, and yanked on her winter boots. With a deep breath, she wrestled open the door, swallowed the brisk London air, tightened her coat, and began her walk. At the corner of our street, a neighbor spotted her and waved eagerly. She waved back. He flashed a smile, pointed at her pregnant belly, and called out:

"You have news!"

She nodded politely and looked away.

"Congratulations! And say hello to Arinze. I haven't seen him in a while."

His name took her breath away. She considered running away but her knees wobbled. She managed to nod, lower her head, and take a few shaky steps back toward home.

She paused at the front gate and again tried to revive the fantasy. She imagined Arinze sitting in the living room helping Obinze with his homework. Perhaps an algebra problem or parallelograms. She squeezed her eyes shut before slowly opening the door.

The room, as usual, was dark and empty.

✌︎

Obinze struggled to readjust to life at school.

My mother had somehow found the strength to notify his head teacher about the accident. In a brief phone call, he echoed familiar lines of condolence.

"We're so sorry. We're here for you. Let us know if we can do anything."

But Obinze had no interest in sharing his pain with his classmates, and in those years, grief counselors in schools weren't common. He fought to project calm, to respond to questions politely, and above all to keep his emotions to himself. He was becoming a man. He needed to be strong for himself and his family, just as he'd been at the funeral.

He learned quickly that his friends wouldn't understand anyway. He bumped into one of them near the pavilion on his first day back.

"Sorry to hear about your dad."

"Thanks."

"I'm off to play football. See you later, mate."

In that instant, a deep well of loneliness exploded in his chest. His dad was gone forever, his mother was locked in her bedroom, and he feared she might leave, too. His classmates only wanted to play a game. And some friends never mentioned anything about his loss at all.

His teachers were more understanding—at least at first. They showered him with warm condolences, gave him extra time for assignments, and offered their time to talk about his grief. Obinze politely declined. He wouldn't—couldn't—open up about the storm of emotions that swirled inside. And soon enough, the warm overtures from his teachers stopped.

It was the same at home. Even when he wanted to talk to our mother about the rage and sadness and hopelessness he felt, he didn't know how. She was hardly in a state to listen anyway.

I, too, was alone that fall as I entered school for the first time in my life—three weeks later than the other kids. My dad had spent the month before his death trying to prepare me. We read books about starting school, practiced tying my shoelaces by the front door and hanging up my coat. We picked out a new backpack and stuffed it with pencils, erasers, and crayons. He explained the school routine, making friends and listening to teachers. He was supposed to be the one to drop me off.

That responsibility, instead, fell to my grandmother, who, though loving in her own way, lacked the gentle emotion my father embraced. When we arrived at school that morning, she simply led me by the hand toward a teacher, thanked them, waved goodbye, and left. There were none of the lingering assurances or hugs the other parents gave their children. Later, I watched bewildered as the other children played together carelessly on the climbing frames and merry-go-rounds. I struggled to understand why the other children seemed happier than me, why I had to spend all day away from home, and why my dad wasn't there to drop me off as promised.

Unaware of our struggles, my mother soon returned to the pharmacy.

She entered through the back door and switched on the lights early one morning. Everything looked the same, the shelves neat and well stocked, but it all felt cold and soulless now. She remembered how much Arinze loved keeping her company here after his medical classes. They'd spend hours sitting between the racks of shampoo and shaving cream, talking about their future. They carefully planned a renovation, more staff, and eventually an expansion. They were always dreaming, my mom and dad.

She thought about the last time she paced these aisles, scrambling for boxes of bandages and gauze in the small hours of the night. How stupid she was, believing she could save anyone with Band-Aids.

She closed her eyes to suppress those memories and tried to focus on the workday ahead. She surveyed the shelves, tidied up the rows of baby lotion and wipes, and scanned her way down, eventually stopping at the row of diapers that needed refilling.

Over the next few days, my mother began to open up to her most loyal customers. They cried shoulder to shoulder as she shared a taste of the grief and sadness that ravaged her—the denial, the struggle to leave her bedroom, and the crushing guilt for leaving her youngest son behind.

The crying sessions with her customers lasted well into the winter. Some would rush in during their lunch breaks to hug her and hold her hand, but there were quiet days, too, days when she couldn't speak to anyone. More than once, in the middle of filling a customer's prescription, she'd politely excuse herself and retreat to the bathroom to cry for as long as it took to feel numb again.

In her pain, she'd sometimes hear hurried footsteps running out of the pharmacy, no doubt shoplifters taking advantage of an unmanned shop. She didn't care. She couldn't care. The most important person in her life was gone. Nothing else mattered.

I saw my mother very little that fall, and Obinze grew more distant as well. At first, he filled his schedule with extracurricular activities to avoid coming home until dinnertime. My big brother eventually started skipping class and somehow found his way to the rougher parts of the neighborhood, where the older kids my mother warned us about offered the acceptance he so badly craved. He spent almost all his time with them.

More than once, he forgot to pick me up from school. I remember waiting alone in the playground one cold afternoon until after it got dark. The school groundskeeper and his wife eventually spotted me crying by the school gates and invited me to wait in their home until my mother got off work. They were horrified and so was my mother when she finally showed up.

Most of the time Obinze didn't tell us where he was, and too often no one was there to ask.

Our family was coming apart.

And then, another call. This time it was a school administrator who explained that Obinze was being expelled. He wasn't showing up for days at a time, and when he did, he was getting into fights. He'd even begun to yell at his teachers. The school had sent several letters home, none of which were answered.

Shaking with disbelief, my mother tried to reason with the man on the other end of the line. She explained the pain that consumed her family, how her son needed support, not expulsion. The next day, she went to the school in person and begged for another chance, but they'd already made up their mind. They would send a formal letter in the coming days.

Dejected, she retreated to her room that night, slammed the door shut, and crawled into bed.

Is this really what's become of my family?

Her husband was gone, her brilliant son Obinze expelled, and Chiwetel lay alone in a dilapidated hospital thousands of miles away.

As she hid under the covers in her darkened bedroom, she was painfully aware that her family's fate was teetering on the brink. She was a single mother raising three children in a neighborhood all too eager to take them away. Yet somehow, with every reason to be afraid, she was suddenly overtaken by a familiar sense of calm as a new thought came upon her.

Remember who you are.

It was a simple, yet powerful mantra for a woman on the verge of surrender.

Remember who you are.

For the first time since that life-shattering phone call, she could see the reality of her situation without fear. Her husband was not coming home. The fate of her children rested with her and her alone.

I now know that the forces that freed my mother from her prison of grief that night were linked to generations of Nigerians who have faced untold pain and suffering and escaped with hope and strength.

That strength was her birthright.

When my mother woke up the next morning, she called for Obinze and me to gather at the top of the staircase. My brother and I slumped down on the cool wooden floor. She fell to her knees and touched us for the first time in weeks, clutching Obinze's hand and kissing my forehead. We melted into her embrace. She promised us, demanded actually, that things would be different from that moment forward. She would enroll Obinze in a new school. She would bring Chiwetel home. She would implement a healthy structure and routine for all of us. She would make life good again. My mother, like countless Nigerian mothers before her, would not let us down.

Chapter 3

Chiwetel came back to us on a cold wet morning in October.
At only eleven years old, he looked world-weary and shattered.
My mother led him through the front door, his right arm still slung in a
cast and his head wrapped in dirty bandages. He whispered something
about the pain in his leg as he walked into the living room. Excited,
I wrapped my arms around his waist tightly, too tightly perhaps. He
winced. My mother called out to Obinze for help, and in an instant, he
was at his brother's side.

Chiwetel was the last person to see Dad alive, the last person to hold
his hand, the last person to hear his voice. For us, he was, at the same
time, a bridge to Dad's spirit and a breathing reminder of his tragic
death. Our father was in the car that day to help Chiwetel. My brother
would carry that weight of crushing guilt for the rest of his life.

Just a day earlier, he lay alone in the middle of a chaotic hospital in
Enugu dreaming of coming home, where the light bounced off the liv-

ing room mantelpiece and ogbono soup bubbled on the stove. He got what he wanted. But home at last, there was little joy.

Obinze barely spoke as he welcomed his little brother home. My mother, who was forcing a smile through her tears, guided Chiwetel gingerly through the house and toward his bedroom in silence.

Years later, my brother would tell me his sudden realization in those first moments home: No matter how close he got to Dad's favorite records, his books, or the games they played together, he would never see him again. Dad was never coming home.

My mother pushed open his bedroom door, flicked on the light, and laid his blue backpack at the foot of the neatly made bed. I watched from the doorway.

Everything was where he'd left it. An interrupted game of checkers sat on the bookshelf and playing cards were scattered by the hamper. He sat on that very same bed playing Whot! with Dad countless times, dancing like fools whenever either of them ran out of cards. That was also where our father would put him to bed with stories of magic and masquerades from our village.

Arinze and Chiwetel were meant to be father and son. They moved through the house, and through life, as teammates. If Dad walked upstairs, his young son followed, skipping up the stairs two at a time to keep up. They were always laughing or joking in one corner of the house or another. Dad's voice was bass to Chiwetel's soprano.

Deeply connected in life, their connection had become immortalized by the circumstances of Dad's death and the deep scars forming on Chiwetel's forehead. My brother slid onto his bed and my mother propped up his head with pillows, tucking another one under his broken arm.

"We have to keep it elevated. Does it still hurt?"

He nodded and turned his head toward the window, where the

last warmth of daylight was slipping away. My mother settled on the spindle-back chair next to the bed. She wanted to apologize—for letting him go on that trip, for leaving him alone in the hospital, and for the pain that was so clearly eating his body and soul. She started to speak, hesitated, then stopped. It was too soon. He'd been home for less than an hour, his eyes barely open, his body weak. There'd be time for all that later.

She reached behind his head and began unwrapping the stained bandage. The protruding scar had formed jagged ridges on his forehead. My brother closed his eyes as she carefully applied ointment.

"I hope I'm not pressing too hard."

A picture of strength on the outside, my mother was struggling on the inside to understand how the youngest person in that vehicle was the only survivor. They said he wouldn't have lived if his right arm hadn't somehow protected his head from the concrete when he was thrown from the vehicle. She imagined the pain and terror Chiwetel must have felt waking up in that hospital alone, how she lied to him for days about his father's fate. The guilt was incredible.

Shaking those thoughts away, she promised herself she'd no longer focus on the horrors of that past September. Good things were happening in her family. Her son could lift a cup of water; he could write again; he was almost ready to return to school. She kissed both temples and said good night.

As much as she wanted to look on the bright side, she still had much to fear. Chiwetel's wounds were far from healed, Obinze was in trouble, I had just started school, and she was five months pregnant and solely responsible for running the family pharmacy six days a week.

Thank God my grandmother was there to watch over us.

As we slept that night, Mama Nnukwu silently unpacked Chiwetel's suitcase and hauled the mass of shirts, pants, and underwear down to the cellar. After only a few weeks in England, she had no idea how to work the washing machine, but after fiddling with the buttons a few times, it clicked to life. The water rained down on her wounded grandson's soiled clothes, and she began to pray, out loud, as she often did. It was a simple prayer: "Thank you for saving our boy."

Upstairs, my mother sat alone at the kitchen table, flicking through a thick stack of paperwork. Shiny brochures of classrooms, semester schedules, and smiling white children singing at recitals and holding trophies. Scanning each prospectus carefully, she feared none of them would accept Obinze after his expulsion. She looked through them anyway, desperate to believe in something.

A year ago, her son could have been mistaken for any of the children in these perfectly staged photos: playful, eager, endlessly curious. Obinze would sulk for hours if Dad was too busy to take him on a guided tour at a museum or to watch war reenactments of the Battle of Hastings. At his first school in Forest Gate, the head teacher once told them that he was too intelligent to be there, that he would be best served at a school that challenged him. Arinze couldn't afford any of the schools he'd recommended so he immediately set about finding scholarships at some of the competitive schools a little farther out. Within a few months, he'd won a place at a reputable boys' school in South London, where he excelled.

But in the weeks after his father's death, it was as if he'd gone into his bedroom one night, fallen asleep, and woken up a different person. Her obedient son was no longer there. She knew he sneaked out of the house at night and was barely there during the day while she was at work. She could only imagine what he was doing out on the streets.

Desperate to keep Obinze occupied as she worked to find him a school, she instructed our grandmother to start watching him closely.

Mama Nnukwu, who rarely had time for idle chitchat, meted out chores in her thick Nigerian accent like a drill sergeant, just as she had done with her own children a generation earlier.

Obinze was stretched out on the sofa watching an episode of *Casualty* one evening when the screen suddenly went dark. Confused, he turned around. Standing at the threshold was Mama Nnukwu, armed with a remote control he didn't know she could use. She pointed upstairs and snapped her fingers, ordering Obinze to bring down the pile of dirty dishes stacked against our mother's bedroom door. Before he could respond, she told him to collect the glut of unopened letters from various tables and counters and place them neatly on her dresser, where our mother could deal with them in peace. Obinze opened his mouth to explain why he couldn't help, but he knew, as we all did, that Mama Nnukwu wasn't interested in excuses.

The tough-love approach took some getting used to, but with her, there was no other way. He nodded reluctantly as she ordered him to peel potatoes, retrieve ingredients from the refrigerator, or stir her simmering stews. And with Chiwetel still broken and bandaged, Obinze was expected to bring his little brother whatever he needed at home—a bowl of soup, a glass of water, or a pain reliever. He also shepherded Chiwetel to every visit to the doctor or hospital. He held doors, hauled bags, sat alone in waiting rooms. It was constant. The pressure on that grieving teenager to be a man, to take care of his family and to live up to his father, was just too much.

One evening, my mother returned from the pharmacy late, cradling three bags of groceries with tired arms. She called out to Obinze for help. Silence. She called out a second time, a bit louder, expecting to hear his footsteps dragging down the wooden stairs. Nothing. She dropped the bags in the hallway as delicately as she could, took a few steps upstairs, and hollered again. Chiwetel and I peeked out from the landing to see what the fuss was about. Suddenly, like a bull, she rushed to the top of the staircase, rapped on Obinze's door, and swung it open. An unmade bed, empty packets of potato chips on the floor, and an overflowing wastebasket. No Obinze. She turned and slammed the door shut as Chiwetel and I scrambled out of the way.

She marched downstairs, quickly moved the groceries into empty cupboards, and, like generations of mothers before her, settled down into a chair by the front door to wait. Her eldest son was still a boy. He had no business being out that late by himself. She sat alone in the dark, arms crossed over her pregnant belly, desperately listening for any clues that cold South London night might have as to his whereabouts. She perked up at every revving engine, every pedestrian's voice, every siren.

In the '80s, South London was making headlines for all the wrong reasons. Newspapers were filled with grim reports of gang violence and drug busts. The local convenience stores—including my mother's pharmacy—were robbed with such regularity it became almost ordinary. Then there were the stabbings, the carjackings, the burglaries.

The violence that already plagued the neighborhood reached a fever pitch in April 1981 when riots broke out in nearby Brixton. Young Black men, tired of having their faces ground against the concrete or their bodies slammed against police vans, swarmed the streets to protest. What started as isolated clashes turned into a full-blown day of reckoning. Television coverage showed teenagers hurling bricks at police, smashing storefront windows, and setting fire to cars over several hours of violent unrest.

And it wasn't long before the dangers crept closer to us. My dad, a young Black trainee doctor, was once dragged into a police van and beaten for no reason other than the obvious. He staggered home that night, bleeding and aching, vowing to file a complaint as soon as he returned from that last trip to Nigeria.

Years after the riots, the streets had only become more dangerous. Pockets of South London were deemed no-go zones. Gangs like the Brixton 28s patrolled the neighborhoods at night and controlled much of the drug dealing in our area. The Summer of '88 was deemed London's "Second Summer of Love" because of the explosion of hard drugs like MDMA that fueled all-night parties.

My mother thought back to the last time she waited with such intensity in this very room. That first Sunday in September, on the same chair, her ankles crossed and arms folded in the same position, waiting for her husband to call. She glanced over at the wall, the same wall she leaned up against to avoid collapsing when the call finally came.

No. This is not happening again.

The scrape of keys twisting in the lock didn't come until almost 6 a.m. Obinze clicked the door shut and tiptoed toward the stairs. His eyebrows lifted in horror as he saw my mother seated quietly in the living room, wearing the same rumpled white blouse and pants she'd worn to work the day before.

Her voice was steady with rage.

"Seven hours. I've been sitting here waiting for you for seven hours."

Obinze was too scared to respond.

"Do you know what I've been through? The worry?"

She swiped a newspaper from the side table and shoved it toward him. "What does it say?"

My brother took a step back, stumbling. Her voice was getting louder.

"Young man murdered on Norwood Road. They still have flowers tied to the lamppost."

She was furious, swallowed up by the kind of rage that consumes your whole body. She started screaming at him—screaming about the dangers outside, screaming about his expulsion, screaming about being on her own.

She grabbed his left wrist with one hand and raised her free hand menacingly. This threat of violence somehow emboldened my brother. He turned his face toward her, unflinching, daring her to strike.

"Go on, then."

Corporal punishment is a way of life in many African homes. Nigerian mothers believe it's the only way to prevent the empty promises, false gods of the Western world from leading their children astray. In our home, if we were ever caught bickering or squabbling among ourselves in public, my mother would immediately cast us a predatory look. Whenever we felt those piercing eyes, we knew it was time to pipe down, sit up straight, and stop whatever we were doing. But that night, those eyes could no longer tame Obinze into submission. He'd already survived the nightmare of his father's death. Nothing could scare him now.

Staring into Obinze's defiant eyes, she slowly lowered her hand. He turned and thundered upstairs, brushing past Chiwetel and me at our viewing post on the landing. She gestured dismissively for us to return to our rooms, her eyes flashing fire. Chiwetel and I exchanged frightened glances before retreating, while our mom scrambled to get ready for work.

It was Friday, and my sleepless mother needed to open the pharmacy in less than an hour. She called up to my grandmother to help make breakfast for us kids before dashing upstairs to change her clothes.

Our entire family stayed on edge that day. The tension hung over us like a thick London fog blotting out the sun. We were afraid to make eye contact when she got home that night. Whenever she walked into a room, we were smart enough to keep quiet.

Just before dinner, she accidentally bumped up against Chiwetel as they passed each other in the dark hallway. When she raised her hand to reach for the light, he ducked and covered his face. After seeing her confrontation with Obinze, Chiwetel was afraid. My mother stepped back and covered her mouth, her hands shaking. Shame burned her insides. There was too much anger in her home, too much strife since Arinze died. If she was going to push her children down the right path, she was going to have to find another way.

Sleep did not come quickly for my mother that night. She was again battling fear. So much fear. As the oldest son in a Nigerian family, she'd expected Obinze to become the public face of the household, to bring pride to the family name. But after his expulsion, his loyalties had shifted to older kids in the neighborhood, kids who'd introduced him to the likes of Marlboro and Absolut. Coming from the village, it was a world she didn't understand. She thought about Obinze's evolution—a young man's body, the hardened expression, the even gaze. Once a sweet boy so eager to please, he was suddenly rebelling against any authority he could find.

And Chiwetel, so sensitive and impressionable, had come home to a family weighed down by friction and pain, even as he struggled with his own trauma.

God, please help us.

Her own mother had raised nine well-disciplined children under far more trying circumstances in Nigeria—a civil war, famine, and extreme poverty, often dealing with all three at the same time. My mother and her siblings were still taught to respect their elders, to take their family responsibilities seriously. Obiajulu had begun caring for her family at the age of seven, feeding her younger siblings eggs and dodo each morning before leading the flock down the bumpy hill to the nearby elementary school. Responsibility was all she had ever known. Of course, she could raise three children here in England. She had no choice.

꙳

My brothers and I started our Sunday the following week in front of the television. The grainy cartoons felt like our only refuge that morning. A thin sheet of snow brightened the cars and trees outside, and the white morning sun made the TV difficult to see, but we stared anyway.

My mother interrupted our morning trance, barging into the room like a sheriff cornering a group of sleeping bandits. She snatched the remote and switched off our show.

"Up. All of you. Dining room. Now."

We followed through the hallway, with Obinze looking particularly bleary-eyed, and stopped at the dining room table, where three books lay neatly stacked. She handed one to each of us. We scanned the covers, then looked at each other as she motioned for us to sit down.

"We've been through a lot this year. We've suffered, yes, but that will not stop us."

She glanced at me but focused on my brothers.

"There will be structure here from now on. There will be routine."

She stared at Obinze.

"I will not let these streets take you from me."

My mother began ticking through a list of changes. Strict curfews for the boys, a long list of chores, weekly reading lists, regular family time to talk academics. There was special emphasis on the reading. From that day forth, we were ordered to read one book a week and discuss our progress at the dinner table. There was no TV allowed at the start. Our eyes were to be glued to words on pages, not programs on screens. For the rest of that day, my brothers retreated to their rooms under her watchful eye, sat at their desks, and studied the new books. She plopped me onto a chair in the kitchen and handed me two short board books to read as she prepared lunch.

The following Friday, the three of us padded downstairs, book in hand, and took our usual seats at the dining table. We slurped my mother's vegetable soup as she tested us, asking us about the story line, the characters, and themes. I was barely in school, but she treated me no differently from my older siblings as she peppered me with questions about the books we'd read together, like *Flat Stanley* and *The Sheep-Pig*. She pressed me to explain Babe's relationship with Fly and the special talents of Stanley, the little boy who could slide under doors.

My brothers were hardly willing participants, that first session. Chiwetel took his cues from his big brother, who laughed off the questions or gave single-word responses. My mother was prepared. Eventually, she grabbed his copy of *The Three Musketeers*, flipped it open, and began reading aloud in her thick-accented English staccato.

> Great criminals bear about them a kind of predestination which makes them surmount all obstacles, which makes them escape all dangers, up to the moment which a wearied Providence has marked as the rock of their impious fortunes.

Families recover from trauma in different ways. For us, this moment, this forced focus on reading—even as one brother bled and the other was schoolless—was the next step in the Ejiofor family comeback. My mother, who spent most of her waking hours at the pharmacy, needed a way to keep my brothers occupied, to give them purpose, to keep their minds centered on anything besides the empty chair at the dinner table.

At home, the reading sessions became part of our weekly routine. If Obinze wasn't actively participating, she'd cut his pocket money for the week and threaten to reduce it more. Then she'd read aloud from his book, as if she could single-handedly will those words to penetrate his teenage defiance and fatherless pain. No matter how long it took, she was going to make reading and academics a dominant focus in her home.

🖎

One morning, she rose early for an appointment at one of the shiny schools from the brochures. She asked my grandmother to fix break-fast and sprinted out the door. After speeding through the streets of Norwood faster than she should have, she scurried into the lobby of a school near our home and greeted the receptionist. The clerk, after commenting on her tardiness, led her through the narrow halls until they reached an office. The door was already open. She shuffled in, shook the silver-haired white man's hand with a forced smile, and sank into the chenille couch. The admissions director scanned his notes from behind his wooden desk.

"Your son would like a place here?"

She nodded.

"And he's starting in the middle of term because . . . ?"

She tried to explain the nightmare of the previous months, that her son was a hardworking kid who'd briefly lost his way.

He peered over his glasses, brows furrowed.

"He was expelled?"

My mother looked down, ashamed.

"He's a brilliant boy. I can show you his reports. He won the top math and science award at school last year, and I promise you he will do better."

"Why don't you try back later in the year? There might be room next fall."

"He needs something now."

"I wish I could be of more help. Call back in June . . . May at the earliest."

My mother had lived in England long enough to understand he was too polite to deliver bad news directly. She stood up, embarrassed, extended her arm for a final handshake, swung her purse over her shoulder, and left without another word.

The dam inside her chest that held all her pain felt like it was about to burst. She'd already received a handful of rejections from some of the good schools in the area. They didn't care how much potential Obinze had or that he'd just lost his father. None of them had room for an expelled child. A puddle of tears sat in her eyes for a minute or so before splashing onto her cheeks. Obiajulu willed herself to have strength. She was going to have to find a way.

*

At home she continued our weekly reading exercises with even more fervor, desperate to instill a sense of routine and discipline in anything, desperate to keep Obinze occupied. It didn't matter that her own relationship with literature was limited to the teenage romance novels she coveted back in Enugu, she was determined to give her children the same comfort, the same opportunities, and the same encouragement they would have had without their sudden loss.

One Saturday, she closed the pharmacy early and made her way to a small bookstore a short drive from our house, arriving just before closing. She was uncomfortable there and it showed. Her eyes darted around the shelves, scanning the racks for titles she recognized. She'd always thought her schooling in Nigeria had given her a decent grasp of English compared to the other children in her village, many of whom never went to school. But now that she stood in a *real* bookshop in England, it dawned on her how little she knew. She remembered her white teachers in Nigeria mentioning a man named Charles Dickens a few times, but she couldn't remember what books he wrote or what they were about.

Summoning all her courage, she asked the store clerk to direct her to the so-called classics she'd heard about from her husband. Arinze loved to read. He was always plopped on the sofa, engrossed in the pages of one novel or another. He loved language, how words could stimulate the senses and create worlds. He would often share what he was reading with his sons, describing in detail a book's complex ideas or testing their comprehension as he read a page aloud from post-colonial African poets like Christopher Okigbo or the short stories of Oscar Wilde.

How she wished he was here to help her with this.

The clerk pulled several books from the shelves: *Frankenstein*, *Lord of the Flies*, and *Jane Eyre*. Mentioning her five-year-old daughter, she asked for the children's section. He marched over to a small corner in the back and picked up Ted Hughes's *The Iron Giant*, Roald Dahl's *The BFG*, and E. B. White's *Charlotte's Web*, and a few other books for children slightly older. Assuming these were titles she probably should have heard of, she was reluctant to give herself away by asking questions. She thanked the clerk politely and left the store, cradling a paper bag packed with new material and promising to be back for more in a few weeks.

Over time, those books and others began to fill our home. At first, my mother relied on the recommendations of friendly bookstore clerks and regular customers at the pharmacy. Later, she discovered newspaper reviews that offered a trove of newer options. After nearly eighteen years in the United Kingdom, she joined a library for the first time. She was usually in too much of a hurry to scan the shelves for more than a few minutes, but if the books seemed to have imaginative plots and slightly more challenging vocabulary, she tucked them under her arm and brought them home.

Friday dinners were focused on literary discussions, our very own family book club. More than once, I caught Mama Nnukwu flashing us a warm smile of encouragement, thankful that her daughter was slowly coming back to life. My mother had barely spoken to anyone in the weeks after my dad's passing. She struggled to leave her bedroom, let alone answer phone calls or open mail. But in our book club, she had found a simple but unifying force to help rally her children together. Something Arinze himself would have cheered.

The discussions and quizzes that followed sparked a rivalry between my brothers, who were separated by just three years. They would blurt out the answers before my mother had even finished asking her questions, each one eager to beat the other in this new game that was holding the family together. Obinze, always a quick study, pumped his fist every time he answered correctly. On some level, we all knew he needed a win. The study sessions around that stained and scratched dining room table were helping him remember who he was.

✍

As much as the book club kept us occupied, it could only do so much. My mother needed to get Obinze back to school. She spent much of her downtime at the pharmacy scanning brochures and making appointments. She wanted a school that would nurture her son's talents; one that carried little risk of him mixing with the wrong crowd and getting into trouble. Her last hope was a small school in Battersea, which, like some of the others, had a reputation for its attentiveness to discipline.

They were invited in for a formal meeting after my mother badgered the admissions office. The school grounds were imposing, but as they passed through the front gates together, the building emanated a sense of warmth. A groundskeeper offered a friendly wave as they walked in. They were ushered to a large office and waited at the threshold for a few seconds before an older man beckoned them in.

The head of year spoke first.

"Your mother told me what happened. I'm sorry about your father."

Obinze, in his freshly pressed jacket and tie, nodded nervously. He opened his mouth to respond but my mother jumped in.

"We've had a difficult time, but we are not here to make excuses. He wants to do better. He will do better."

"Why do you think it'll be different here?"

"We sent him back to school too soon. He needed time. He's had that now."

The teacher scanned Obinze's school reports.

"You've excelled in the sciences, I see. And your math scores have been consistently high. But we don't usually accept students in the middle of term. And the expulsion makes us—"

My mother cut in.

"My son has legitimate potential. All he does is read—Kipling, Austen, Hardy." She knew it was an impressive list, even though she'd only just begun to learn their names. She looked straight into the teacher's eyes. "He just needs guidance, someone to steer him."

He scribbled down a few notes.

"He'll have exams to sit, then interviews, of course. I can't make any promises."

"We're ready."

"I understand. We'll be in touch."

My mother left the office with Obinze in tow, not sure whether to trust that glimmer of hope she sensed in her chest.

"I did okay, didn't I?" Obinze asked, glancing up for some kind of reassurance.

My mother gently rubbed his back. "I think so, darling."

"I wish things weren't always so hard," he said.

He reached out to her with both arms, and she welcomed him in, squeezing his broadening shoulders. It was one of the first embraces they'd shared since the funeral and the first time he'd lowered his hardened teenage exterior since they'd returned from Nigeria. She pulled away after an extra moment, gently wiped her eyes with the back of her hand, and climbed into the car.

That evening we sat at the dinner table, reviewing the latest texts we'd been asked to read. My mother fired a round of math questions at Obinze to prepare him for his upcoming tests. The brochure for the school in Battersea she'd just visited was sitting on the bookshelf in the corner. She glanced at it every so often, praying silently to herself. It was the jump start she needed to get her boy back on track.

⟳

The following week, my mother busied herself in her pharmacy, sticking price tags on hairbrushes and detergent.

Friday afternoons were usually the most frantic. A flood of customers often rushed in on their lunch breaks to pick up prescriptions before the weekend. She had an hour left before the frenzy started. She took a deep breath and savored it. She was in the middle of this stolen moment of peace when the phone rang. She leaped up, headed behind the counter, and picked up the receiver.

"Mrs. Ejiofor?" a female voice said.

She assumed it was a local doctor calling in a prescription and slid a notepad and pen toward her. "Yes?"

"This is Jennifer Deacon. Thank you for applying to our school." My mother held her breath.

"We're calling to offer your son a place starting next week."

She gasped, covering her mouth in disbelief.

"Thank you, thank you," she responded, fighting back tears.

"We'll send you the information you need in the coming days."

She hung up the phone, chuckling with joy, her face beaming brightly for the first time in forever.

There was still a speck of surprise in her voice when she told Obinze the news. He asked questions; she answered excitedly. Soon they relaxed into a shared sense of satisfaction, something they hadn't felt in a long time.

✿

Christmas was colder that year, more muted. It was our first without Dad, our first without him wrapping his arms around us, singing along to "Jingle Bells" in a booming Nigerian accent. My mother woke us up early and led us downstairs, where we lined up along the living room wall, quietly examining all the gifts she'd stuffed under the Christmas tree—and her not-so-subtle efforts to include our father. There was a new photo of him on the mantelpiece, and one of his songs was playing softly on the cassette player.

It was not the kind of Christmas my father would have approved of. He always tried to re-create the raucous Christmases of his Nigerian childhood, the kind of holiday that felt more like a crowded street fair than a solemn church service. He found it surprising that Christmases in England were so quiet, with families spending the morning home alone among empty streets and closed storefronts. So, Dad went out of his way to do things differently. He usually invited three or four other large families to celebrate Christmas Day in our home. He made sure there was no shortage of song, dance, or drink, and at some point he always ended up performing atilogwu as the kids cheered.

We accepted that Christmas that year was going to be different. With the photo of our father looking on, the three of us converged on the Christmas tree to retrieve our gifts. I squealed with delight when I opened my new Etch A Sketch, and my brothers found their own moments of joy when they opened their big gifts.

Once the frenzy was over, my mother motioned for us to join her on the couch. We snuggled up together, giddy from our new spoils. My mother was smiling, too, as she gently lifted my hand and placed it on her belly.

"You have one more gift," she told us. "You're going to have a sister."

Chapter 4

Nigerian names tell a story. In a word or phrase, they often represent a family's aspirations, a vision for the gifts their new child will someday bestow upon their communities. *Eniyii, a child of integrity; Titilayo, a child who shines with happiness.* They also encapsulate the family's emotions at the time of birth. Some names require entire sentences to be translated from our native Igbo language into English. *Chiwetel, he who is brought by God. Obiajulu, my heart is at peace.*

My little sister came into the world on a Tuesday morning in early spring, and in those opening minutes of her life, she did not have a name. My mother was alone that night, save for the doctors and nurses, in a cramped hospital room as the three of us older kids slept at home under my grandmother's watchful eye. Childbirth was not a new experience for Obiajulu, but giving birth without the support of her husband was.

She was exhausted, spread-eagled on the hospital bed, when a nurse gently placed her infant daughter on her chest for the first time.

Her tiny head fit perfectly into the palm of my mother's right hand. Despite the turbulence of the past six months, she'd somehow arrived in this world safely. A clutter of emotions—hope and regret, love and loss—stirred within my mother as she gazed at the baby's sleeping eyes. This helpless being was the latest and last breathing evidence of her union with my dad—the final punctuation of a decades-long story of togetherness. She looked at that baby girl like the offering from God she was, the Lord's proof of life after death.

The last time my mother lay in this very hospital was for my birth, and my dad was at her side. He leaned in close to her, nervously squeezing her hand during the labor, sprinkling just the right number of technical terms in his conversation with the obstetrician to make it obvious he, too, was a doctor.

But she was alone on this night when the nurse asked for the child's name for the first time. She hadn't settled on anything specific ahead of time, preferring to meet her fourth-born to determine what felt right. Still, it took only a few moments for my mother to respond.

Kandibe.

It was an Igbo name that loosely translated to, *Despite what I've been through, I will keep going.* She whispered the name to herself and to her sleeping child a few times to make sure it fit, before nodding assuredly and declaring her daughter's name to the nurse—and to the world.

Despite what I've been through, I will keep going.

꒰

The next morning, my mother carefully laid Kandibe, or Kandi as we'd later call her, into her basket and shuffled out of the room toward the hospital's double doors. It was a cool, rain-soaked morning. After she stepped outside, she knelt down and cocooned the baby tighter in the wrap before being satisfied she was warm enough for the short ride home. Scanning the street, her eyes landed on a taxi driver. They nodded at each other and he pulled around. Obiajulu, too shattered to speak, was staring down at her daughter the entire drive, searching her scrunched up face with all her might for any sign of Arinze. He was there. He had to be there.

Obinze, Chiwetel, and I were vibrating with excitement when the tap dance of my mother's keys shook the front door open and we could at last meet the new addition to our family. We jostled, pushed, and elbowed our way to the basket, our eyes wide with wonder. We inhaled her tiny features and serenaded her with a chorus of loving *awwws* as my mother tried to defend her delicate bundle.

My grandmother shooed us out of the way to clear a path to the stairs for my mother, who was beyond exhausted after several hours of labor and barely any rest since. She lifted the basket with a tired grunt and waddled toward the staircase, her belly still swollen and tender. We followed slowly behind her, peppering her with questions as we moved upstairs in unison, desperate to be close to this tiny sleeping symbol of hope we so badly needed.

We were entranced by our new sibling. My brothers and I sat on my mother's bed, consumed by this beautiful being who'd appeared from nowhere and promised to reshape our lives for the better. Aside from the blaring horns and traffic outside, our house had been enveloped by a penetrating silence for the past six months, but now we had a new soundtrack: Kandibe's coos and cries of innocence.

My mother went back to work two weeks after giving birth.

She yearned to stay home for those delicate early months, when every whimper, gurgle, and grimace felt important, but every day she wasn't refilling prescriptions behind the pharmacy counter she was losing money—money she needed for the mortgage and for repairs around the house: the temperamental boiler, the cracked kitchen window, the leaky faucet.

Before leaving for work that first day, she reeled off a long list of instructions to her own mother, who'd raised nine children and certainly didn't need them but listened intently all the same. She arrived at the pharmacy guilt-ridden and resentful, exhaling deeply as she unrolled the metal shutters and slipped in through the back door. She carefully placed her bag on top of the cubbyhole shelves, each square an alphabetical depository for prescription medication. At work, she had the order and organization that her life at home did not.

Those first few weeks and months back were hard. After long nights with Kandi she'd stand behind the counter for ten-hour days on wobbly legs, greeting customers with zombie eyes as she counted down the minutes to when she could return to her children.

At night, she tried to sleep during the short stretches when the baby slept, clenching her eyes shut as though sleep were something you could beat into submission. Despite the bone exhaustion, she'd usually lie awake worrying into the early hours, even when the streets grew quiet.

In those early days, she tried hard to keep our routine. Just before Easter break, she told my brothers to ask for their teachers' home addresses so they could mail them practice tests over the holidays—a makeshift correspondence course that made them too busy to get into trouble. Some of the teachers actually graded the mock tests and mailed them back. My mother reviewed them carefully, checking them against a printout of the answers and going over what they got wrong.

At dinner she'd press my brothers about their main subjects— English, math, history—desperate to ensure they were staying focused.

She'd ask Obinze what he was learning about the Industrial Revolution, while giving Chiwetel simple equations to solve as he ate.

With her attention mostly trained on my brothers, I often ate my rice and beans in silence, watching my bigger siblings answer her rapid-fire questions. I yearned to impress her, too, to show off how much I knew about *James and the Giant Peach* and the big new words I'd just learned how to spell. But there was only so much attention to go around. Sometimes, while I was reading aloud, my mother would stop me, look up toward the ceiling, and listen intently. Then we'd hear it: the faint screams of my baby sister upstairs. She'd race to the bedroom to tend to Kandi. And that night's book club was over.

The financial strain was also taking a toll. The pharmacy, which was in a neighborhood devastated by unemployment, was struggling to generate the sort of business she needed. There were days only a handful of customers trickled in. She couldn't afford both heat and the expense of raising four children, so, when the weather turned cold outside, it was cold inside, too. It was so cold, in fact, that we could sometimes see our breath in the evening when we spoke. We learned to do our homework in bed under the covers, the only place it felt warm. The colder it was, the faster we'd rush through our assignments. We'd then find comfort next to the only space heater in the house, which happened to be in front of the living room television.

Most days, my brothers took turns walking me home from school. But getting home didn't mean we could relax. More than once, Chiwetel and I made the trek only to find leaves and dirt under the mat where the faded spare front door key was supposed to be, and no one inside to let us in. We were locked out, the key misplaced during the mad morning scramble or my grandmother's long walks with Kandi. Not knowing when Mama Nnukwu would return and hours before the end of my mother's workday, we'd go door-to-door, knocking on homes along our street with frozen hands to find shelter.

We didn't know many of our neighbors, and trust was in short supply in our South London neighborhood. Nervous mothers would peek out their doors and listen to our pleas with skepticism, fearing it was some kind of elaborate ruse. I vaguely remember strained conversations on doorsteps, as Chiwetel tried to explain that he and his little sister needed to wait in their living room or kitchen "just for a few minutes" until our mother got home. As soon as we found a warm place to wait, we'd watch TV and play with their pets for hours sometimes before our mother would appear, frantically apologizing to the neighbor she hardly knew, before dragging us home for dinner.

I now know that my mother was worried she was failing us, drowning in an endless cycle of doing and rushing and earning. People outside our neighborhood began to notice, too.

✒

My elementary school was set along a tree-lined street in suburban Dulwich, a short drive from the manicured gardens and ponds of the park for which the neighborhood was named. But that peace didn't extend to the inside of my classrooms, where my schoolmates and I usually spent more time throwing pencils and drawing things we shouldn't than learning. The teachers, most of them used to such behavior from the younger children, often ignored the disruptions, hoping they'd fade for lack of attention. They didn't.

I was almost seven and, amid the chaos, struggling to understand how I fit into my new world. At home, I was suddenly a middle child, no longer the center of attention. And my father was gone. But it was at school where I felt the most alone. I didn't know how to connect with all those laughing little boys and girls who seemed to have such perfect lives. I was the only Black child in my grade and too self-conscious to speak up in class. I often walked through the halls, withdrawn, desperate to swap my stained and wrinkled uniform for the other girls' pristine shirts and skirts. My longing eventually turned me into a thief.

Once after gym class, I reached into another girl's open locker and traded my unsightly shirt for her well-pressed blouse. I told myself it would look like an innocent mistake if I got caught. Well, I did get caught. The teacher quickly ordered me to give the other girl back her clothes and sent me to see the headmistress.

After school was worse. Standing alone on the edge of the playground waiting for my brother, I'd watch parents come for the other kids—a mother crouching down to kiss her young son on the forehead, a father scooping up a pink backpack in one arm and his daughter in the other. I watched with a deep longing, usually until the playground was empty. I wanted to be a part of those families—families that had two parents, families that picked up their children on time, families that walked home hand-in-hand and had normal conversations about their days.

These feelings continued for more than a year. And it probably would have gone on longer if not for the letter my mother found one night in the crowded letterbox bearing my school insignia. She peeled it open and quickly scanned the first sentences. Her heart sank. She'd missed the last few parent-teacher meetings. The headmistress was urging her to come to the next one. She didn't say exactly why, but clearly, it was not good.

She knew she hadn't been there for me like she wanted to for quite some time. When Dad was around, we used to spend rainy Sunday afternoons in the living room building forts with the couch cushions. And when the sun came, we'd take long walks in Regent's

Park, sometimes going for boat rides on the small lake. She always let me sit on her lap.

But after the accident, there wasn't much time for fun and games anymore. Although she took my brothers' reading assignments seriously, too often she simply handed me a blank sheet of paper and some crayons to keep me occupied. And then there was Kandi, whose arrival had consumed most of the time we would have spent together.

The parent-teacher meeting was set for the following Tuesday. She closed the pharmacy a few minutes early that night, apologizing to the lone customer scanning the store's only two aisles as she gently ushered him out the door. She jumped into her lime-green Honda Civic and raced through the streets of South London, honking at the slow-moving cars and yelling out the window like she was still in Enugu. The gritty commercial hub of Brixton soon gave way to the quiet residential streets of West Dulwich, where my brightly lit school reminded her of everything she was doing wrong.

She burst in through the double doors looking flustered. It wasn't the first time she was late for a school function, but she was more panicked than usual when she arrived in the large hall. There were clusters of white parents and teachers dotted around the room. Smiles, handshakes, and lots of nodding as the parents lapped up effusive compliments about their little ones from the teachers. My mother, not sure whom to talk to, was afraid. Her stomach did cartwheels when she saw my head teacher, who seemed slightly surprised to see her, as though she hadn't expected her to show. She beckoned my mother over with a short wave.

More than a year had passed since my father's death. Still, the teacher opened the conversation by offering her condolences, explaining it was normal for family trauma to take a toll on children. My mother nodded along, but her artificial smile and widening eyes were screaming one message above all else: *Please don't tell me I'm a failure.*

My teacher did not get the message. She dove into her concerns about my poor performance, the missing homework, the sullen disposition, the struggle to make friends. My mother scanned the teacher's face for some sign of reassurance, some sign that her daughter would recover, but none came. For a moment, she saw herself through the eyes of the villagers back home, whom she had left for the sole purpose of giving her children a better education. The thought alone made her shudder. When it was her turn to speak, she tried to make a few excuses before reluctantly admitting she had work to do.

After speaking with a few other teachers—many with similar concerns—she muttered an excuse to leave early. She smiled through the shame and was moving toward the door when a flicker of an idea stopped her. She turned, marched back to where my teacher was standing talking to another parent. She waited her turn and then, with all the confidence she could muster, blurted:

"I'd like to see my daughter's school syllabus for the rest of the year."

The head teacher seemed to want to ask why but changed her mind. She walked over to a closet in the classroom to retrieve a thick folder. Sifting through, she pulled out a stack of stapled papers, forced a smile, and held them out to my mother.

The sky was a tablecloth of dark gray and a stiff breeze was shaking the windows the next evening when my mother came home from the pharmacy. After a quick dinner, she went upstairs to rock her budding toddler on her shoulder for a few minutes, humming the same Nigerian lullaby her mother had sung to her. And once those eyes closed, she laid the peaceful child down in the crib and kissed her good night.

She slipped out, carefully closing the door, and headed straight for my room carrying the small stack of stapled paper. She wanted to talk. Some of my teachers were concerned, she explained matter-of-factly. She wanted to understand what was happening at school. *Did I fit in? Was the work hard? Did I understand the homework?*

I didn't know what to say. I responded with empty yesses and nos, looking up at her with sorry, sad eyes. I knew I had let her down. I knew how hard life without my dad was. I desperately wanted to make it easier.

She apologized for not spending more time with me and asked what I wanted to be when I grew up.

"I want to be a doctor like daddy. Or a postman."

She smiled, which made me smile.

"That's why we're going to work harder in school. You have an opportunity here that we never had. We cannot waste it."

She presented the stack of paper with a serious look.

"These are the subjects you'll be learning in school this year. We'll be learning them ahead of time, together."

She was going to do something her mother had done for her a generation earlier, something countless Nigerian mothers have done for their children. She would teach me what she could of all my subjects months before they were presented in class. Her plan was to give me such a head start that by the time something new came up in school, I would have already mastered it. This was going to be our new routine, and really, it was the only answer she had to solve my academic problems.

My mother—who worked six days a week—spent most of her waking hours on her feet at the pharmacy. She didn't know how to get me to enjoy school or to develop more confidence in class. But she knew that if I started doing better, then maybe, just maybe, the other things would fall into place.

She dragged an old chair from the dining room up the stairs to my room, slid it next to mine, and sat with me for hours that first night going through my times tables and reading comprehension.

The second night, she staggered into the house late after work, teetering as usual with exhaustion. She hung up her coat and pulled out a bowl of leftover egusi soup and plantains from the fridge. She couldn't quite remember how many days old they were, but they'd have to do. After dinner, she came into my room, settled down her pens and notepad on my desk, and went back into the hallway. She unhooked the big white clock from a nail in the wall and brought it back to my room. We were going to learn to tell time.

"This is how we know when to go to bed, when to wake up," she explained.

She guided the hour hand toward each big number, explaining the term "o'clock," then moved quickly on to the minute hand. We spent what felt like hours in front of that big clock on my desk as she taught me the basics. I wanted so badly to show her I understood, but I couldn't grasp the notion of time right away. I'd heard my brothers mention leaving somewhere at, say, "five o'clock" or needing to wake up at "seven o'clock," but I never understood where those magic numbers came from or what they had to do with the big numbers on the edges of the white circle. I got most of my mother's questions wrong that evening but we kept going that night, and the next night and the night after that, until I'd mastered it.

This was how our time together would go in those defining years. She eventually began buying herself additional copies of my textbooks to go over the lessons planned for the coming months. Some nights we'd be up until 1 a.m. That's far too late for a seven-year-old but some-how I knew that whatever we were doing at night was changing my experience at school for the better.

It began to pay off almost immediately. I remember going to class one day after we'd cemented our nighttime routine into law. It was a Monday morning and Mrs. Rossetti, our math teacher, turned to the class and announced we were going to learn our times tables. As she began drawing up the two times table on the blackboard, I lit up inside, knowing we'd already covered my twelve times table at home weeks ago.

"If two times five is ten, what is two times six? Who knows the answer?"

After a few wrong answers from the other students, she eventually came to me.

"It's twelve, Mrs. Rossetti," I said with a smile.

She seemed somewhat taken aback.

"Yes. Quite right. Twelve," she said, her eyes lingering on me an extra beat.

This simple moment marked a major turning point in my school career. I came alive in school. My confidence surged. My teachers treated me as a role model of sorts for the other pupils. School shifted from being a scary place, a place where I never felt comfortable, to the one place where I knew I could shine. I felt a real sense of purpose that fueled my desire to do even better.

At home, I began to love our routine. I'd hear the jingle of my mother's keys announcing her arrival every evening at 7:30. I'd rush upstairs to prepare for our appointment, pulling my worn exercise books from my satchel and neatly stacking them on the desk. She'd prepare dinner, guide the boys through the family book club, put Kandi to bed, and come to my room and begin to teach.

As I got older, I realized our study sessions also gave my mother new purpose. She'd been sucked into grief after the accident, unable to escape the dark reality of her new life for longer than she'll ever admit. Kandi's birth began to break that spell, and our lessons helped shine that light brighter still. Instead of retiring each night to a bedroom cluttered with memories of her lost husband—the bed they'd shared for more than a decade, the framed photos of better times—she now spent most of her evenings in my room. My textbooks, equations, and comprehension exercises became a welcome distraction from a pain-filled past.

It wasn't long before my head teacher noticed something in me had shifted.

One afternoon, she pulled me aside in the hallway and offered real words of encouragement.

"I'm impressed by your improvement. Keep doing whatever you're doing."

By that first summer, I was winning scholastic awards, earning gold stars, and being honored in front of my classmates. My head teacher intercepted my mother as she dropped me off one morning before school.

"She's been so involved, so enthusiastic. What's changed?"

"She finally feels she belongs," my mother responded.

The positive feedback reinforced my mother's commitment. Even though she was now playing the role of father and mother—breadwinner, nurturer, disciplinarian, cook, cleaner—she did not relent. As backbreaking as the pace was, she was desperate to ensure our shared tragedy would not define our future.

☙

My academic performance was soaring but things at the pharmacy were not. Britain had entered a recession, and it felt like our neighborhood was its epicenter. The boom of the late '8os had given way to collapsing housing prices, growing bankruptcies, and surging unemployment lines. Rising taxes led to riots. Friends and neighbors lost homes, storefronts near my mother's pharmacy closed, and crime exploded.

My mother had her usual concerns about the shop inventory and mortgage payments but her safety at work became a daily preoccupation. Burglaries and car break-ins around the council estates that bordered her pharmacy grew more frequent. Clashes with police plagued the neighborhood. Police officers sometimes stopped by the pharmacy to ask if she recognized pictures of suspected criminals. She pretended she didn't, even when she did.

Despite her precautions, she was alone, as usual, one afternoon when they came. Four teenage boys entered the small shop, waited for the only customer to leave, and blocked the entrance. Two of them charged toward my mother, demanding she empty the cash register—or else.

Her heart fired like an automatic weapon. For a few seconds, she seriously considered fighting back. She could barely afford raising four children as it was—a robbery like this would be devastating to her family's finances. She locked eyes with one of the boys, and seeing his violent desperation, quickly unlocked the register and spilled the notes onto the counter. She had young children at home who could not afford to lose another parent.

My mother didn't mention the robbery when she came home. And even that night, we still studied. She treated our study time as seriously as she would any paid job, interrupting our sessions only to tend to the crying baby.

The nights were long and arduous, but once in a while we'd slip into laughter. I would always notice, because she wasn't someone who usually had time for levity in those days. Her laugh was like birdsong, warm and mysterious, but it never lasted long. She'd quickly catch herself and redouble her focus on the material at hand.

My mother's intense approach was shaped by her own desperate experience with education.

Obiajulu Ejiofor came of age during one of Africa's most bitter and deadly wars. The Nigerian Civil War of the late 1960s, known to us as the Biafra war, ultimately left an estimated 2 million dead—most of them from starvation—and sent families like my mother's scattered across the country seeking food and safety.

My mother's clan, even with seven children at the time, was constantly on the move in those years. They found temporary refuge in various towns across the east, trading lodgings every few months while sometimes living in fear behind enemy lines.

Most of the schools were shuttered. The only chance for kids to learn came at home. And my grandmother became a neighborhood teacher of sorts. While she had minimal education of her own, I'm told that she would regularly gather several young girls wherever they were camped, shake out her mat, sit, and teach them whatever she thought they would be learning in school.

The lessons would often take place in the bush, far away from patrolling soldiers. The girls, many on the edge of starvation, would sit cross-legged for hours at a time as my grandmother told Igbo folk stories and pored over basic math with only dusty ground to write on. She focused mostly on the younger children and gave the older children donated books to work through on their own. She created makeshift classrooms when she could, using stools cut from African bamboo stalks, to make it feel more official. And like my mother, my grandmother held classes no matter what.

The dedication to learning paid off. When the war ended in 1970, and my mother returned to school after a three-year hiatus, the gaps in her knowledge weren't as wide as they were for some of the other children.

She earned the highest honors in her high school exams that year, the first time a woman from her village had ever achieved such distinction. Her accomplishment befuddled almost everyone who knew her. *How could a girl who missed three years of school do so well?*

My mother's sudden commitment to our academic futures was decades in the making. Her extraordinary drive to teach me as she had been taught was life-changing. With each math problem we solved, each perfect sentence we crafted, we were writing a better future. I gave it everything I had because she was so obviously giving it everything she had.

Her lessons extended far beyond the classroom. My life, and the lives of my siblings, would later hinge on this simple ritual of learning in advance, this commitment to getting ahead even when it wasn't expected.

And with that, a new world began to open up to us.

Chapter 5

My father was a man who loved music well before he loved my mother. Arinze Ejiofor fell hard during the Beatlemania craze of the early '60s, and he quickly established himself as the front man of a high school band modeled after Liverpool's finest. Of course, he fashioned himself as the Paul McCartney of his group, writing many of the songs and playing guitar. And like any good teenage musician, he spent far more time studying his idols' lyrical genius than his schoolwork.

He was also a dancer. During school holidays, my father would spend hours performing traditional atilogwu in the village square as smiling neighbors clapped along. It was as if he could fly, his feet pulsating to the beat of the drums, his forehead glistening with sweat. Atilogwu is a traditional Nigerian dance known for its speed and beauty that has been celebrated in our villages for centuries. In English, the Igbo word *atilogwu* loosely translates as "this must be magic." Villagers believe the dance steps are so incomprehensibly quick, they can't possibly be the work of mere mortals without supernatural assistance.

As difficult as it was for most people, atilogwu came naturally to my father. Even after the war started, at least early on, he performed his dances for the solemn crowd of villagers in Oghe, giving the hungry and hopeless a reason to smile. Before they were forced to flee, Arinze and his dance troupe spent their evenings hammering away the town's worries and sorrows on their drums. Whenever the beat reached a racing crescendo, grandmothers stood up and cheered and village elders remembered the days before the war, when life was simple.

Music and dance offered my father emotional refuge from war and famine, and maybe more importantly, they fed his soul. But like so many good things in those years, it didn't last long.

Years later in London, after the Biafra war had come and gone, after his own children had begun to arrive, and after my mother had convinced him to focus on medicine, he did not give up on his passion. The arts became a constant preoccupation that followed him into each day's free moments, whenever they happened to occur. He tinkered with songwriting on the train, devoured literature in bed as my mother slept, danced a few steps of atilogwu while celebrating his young sons' triumphs. He had the talent and he had the desire. But in the end, he and my mother agreed that he would devote his final years to a more stable profession like medicine.

🙠

And so, perhaps none of us should have been surprised by Chiwetel's passion for the arts. My father's creative flair, and my mother's insistence on literature, especially the classics, provided the perfect breeding ground for his early interest in plays.

He stumbled upon the arts as a young high school student almost by accident, drawn in not by song or dance but by a sixteenth-century playwright whose prose spoke to his soul. It was a connection he so badly needed.

Up to that point, my brother's two years in school after Dad's death were a study in loneliness. Chiwetel kept his head down as he walked to school alone those first days back after the accident, uncomfortably aware of the swollen, jagged scar on his forehead. The other boys, incapable of understanding the emotional earthquake that had shaken his life, pointed and whispered as they passed him in the hallways. Still, he fulfilled his responsibilities well. He handed in homework on time, participated in class experiments, and played soccer. But a piece of him—the most important piece—was detached, from his classmates and from himself. There was an emptiness in his eyes. He struggled to care deeply about anything.

And though just thirteen, his duties at home were more urgent than ever, especially after my grandmother returned to Nigeria. When Obinze wouldn't come home some nights, Chiwetel, who was barely a teenager, would join my mother for long drives in the early hours searching the neighborhoods for him. And on top of schoolwork, he was still required to read one book a week for my mother's dinner-table book club.

I didn't know it then, but the books weren't a chore for him at all. Once the weekly reading lists no longer challenged us, my mother experimented with seeing if her sons could read a book in an even shorter timeframe. More than once, she handed Chiwetel a book before heading to the pharmacy on a Saturday and asked him to read it from cover to cover by the time she returned that evening.

We all have natural gifts. Some people can shoot a basketball, some can sing, others can play piano. God gave Chiwetel the ability to read fast, really fast, and through practice, the capacity to digest the deep thoughts and themes of authors like George Orwell or Ted Hughes at an early age.

Some of these gifts came from my dad. Some evenings, after Dad came home from shifts at the hospital, he'd often set aside time with his sons to listen to music in the living room. Dad was a doctor but, in his heart, also an artist with an unusually curious mind. He appreciated the way songwriters used language to convey emotion. Bob Dylan was one of his favorites. He wanted his sons to hear beyond the singsongy melodies and into the depth and wisdom of the artist's prose.

They say ev'ry man needs protection
They say ev'ry man must fall
Yet I swear I see my reflection
Some place so high above this wall.

Chiwetel adored Dad and yearned to be like him in every which way. I suppose that's most boys' fate. But it was especially strong with Chiwetel. He would steal these moments to strum my father's old guitar, study pages from books he read, and dance to his music. He lit up when my dad shared stories of his dancing in the village, his old bands, his creativity.

Whenever Dad handed Chiwetel the lyrics to a song he was working on for a later discussion, Chiwetel's entire day revolved around it. He memorized the lyrics, carefully made notes for the discussions that would inevitably follow, and offered up ideas that were incredibly profound for an eleven-year-old.

But their conversations about songwriting sometimes morphed into poetry and literature. The two were bonding over fiction even in their final moments together. My dad had given him a copy of Robert Graves's *I, Claudius* just as their Nigerian road trip began. The book was in Chiwetel's hands on the road to Lagos when the unthinkable happened.

My mother would later turn those literary discussions into more formal weekly reading lists that lacked some of the excitement and wonder my father had injected but gave our family the direction we so needed at the time.

By the time Chiwetel started high school, his favorite subject was English. Those forty-five minutes each day were a glimmer of color in a gray existence. And so it was in one of those English classes, on an otherwise unremarkable Wednesday morning in November, that he came upon the moment my parents had spent their entire lives preparing him for.

That morning, he trudged alongside the other boys down the long, wet path through his school's black wrought-iron gates and into the ornate brick building, scraping a muddy path directly into his seventh-grade English class. Chiwetel settled himself in his regular seat in the third row as the teacher marched down the aisles handing out copies of the class's next assignment. It was a play. They had typically focused on books like Tolkien's *Lord of the Rings* and Thomas Hardy's *Jude the Obscure*, but on this day, they were graduating to Shakespeare's *Henry IV, Part 1*.

The teacher called out each student's name and their corresponding part in the play. Chiwetel was to read Hal, the king's complicated son. The class began slowly at first, stopping every few paragraphs to discuss the complex prose. Eventually, it was Chiwetel's turn. As it turned out, thirteen years of appreciation for the written word from his mother, and an understanding of the power of poetry and creativity from his father, had been building to this one moment in the fall of 1990.

Chiwetel turned to the end of Act 1, Scene 2, where Hal, the future king, addresses his companion, Sir John Falstaff. He cleared his throat and read aloud with a passion that stopped the teacher cold in his tracks.

And like bright metal on a sullen ground,
My reformation, glitt'ring o'er my fault,
Shall show more goodly and attract more eyes

Than that which hath no foil to set it off.
I'll so offend, to make offense a skill;
Redeeming time when men think least I will.

The entire class, the teacher among them, fell silent. After a few moments, he asked Chiwetel to explain what he'd read.

"The prince is telling us that his fall from grace is part of a plan. He's saying he'll eventually return to glory and shine brighter than he ever could have before because people won't expect it."

The teacher, mouth slightly ajar, nodded, not quite believing how quickly and effortlessly this young boy had deciphered the text.

After class, my brother found himself in the library searching the shelves for the Bard's work. He fingered a copy of the same play he'd just read and turned to the same monologue that called to him during class. In the play, Hal, the prince, knew the world had low expectations of him—in fact, he deliberately helped lower them—but one day he was going to shock them all with his success; he might have nothing now, but one day he would be king. Chiwetel felt nothing like a king in that moment, but like his protagonist, he knew he had more to offer than he had been willing to share with the world so far.

My brother marched through the school gates, suddenly and profoundly moved after discovering himself in the tragedy and drama Shakespeare had so beautifully crafted four centuries earlier.

✍

My mother had no idea about her youngest son's new obsession. And maybe it was better that way. The performing arts didn't put food on the table where she was from, they didn't bring money or pride or prestige to the village. And so, to her, music, poetry, and plays were mostly irrelevant—at best, a nice way to pass a bus ride, and at worst, a serious distraction from far more important scholastic endeavors.

She could be forgiven for not noticing right away that Chiwetel was coming back to life. Each day was a marathon of maternal responsibility: get the older kids to school and Kandi to daycare, ten hours on her feet at the pharmacy, cook dinner, put the baby to bed, study with me, and keep up with reading assignments for the boys. If she had any extra time, she gave it to Obinze. He was off to a better start at his new school, where teachers used words like *potential* and *natural ability* on his report cards. But she knew the allure of the streets was an omnipresent threat.

She was less concerned about Chiwetel. She'd once heard something about him participating in a play after school, but to her, it was no different than Obinze practicing fencing or soccer in his spare time. She was sure her sons knew, just as she did, that following in their father's footsteps and becoming doctors was not only the smart choice, it was the only choice. Chiwetel was lucky enough to receive financial assistance from his school, and she couldn't let him throw it away.

In the coming months, Chiwetel would devour the Bard's writings with all the fervor that our father had devoted to the Beatles a generation earlier: *Macbeth, Richard III,* and *Othello* were my brother's "Twist and Shout," "Strawberry Fields Forever," and "All You Need Is Love."

↝

My brother auditioned for his first theatrical performance after school one Thursday without telling anyone in the family. He arranged for Obinze to pick me up that day, making up something about a class project. But really, he was secretly plotting to score a major role in *Measure for Measure.*

Like many of Shakespeare's most famous plays, he knew *Measure for Measure* inside out by that time. He had no acting experience to speak of, and certainly no training to help him prepare, but he auditioned for the part of Angelo, one of the central characters. He spent hours and hours memorizing the lines, and it wasn't work for him. His world already revolved around Shakespeare.

He was the last to audition that day. Even though he should have been nervous, his reverence for the playwright helped mask any fear as he faced a small panel of teachers from an empty stage. He delivered his lines with a passion and presence that caught his audience by surprise.

> By so receiving a dishonored life
> With ransom of such shame. Would yet he had liv'd!
> Alack! when once our grace we have forgot,
> Nothing goes right: we would, and we would not.

The teachers exchanged knowing glances. The school had found its new star.

Rehearsals began almost immediately. My mother still didn't know about the play, but she learned of my brother's uncanny devotion to Shakespeare a few nights later after walking into his room to see several lines from his play scribbled in black marker all over his blue walls.

> Our doubts are traitors,
> And make us lose the good we oft might win,
> By fearing to attempt.

> Thyself and thy belongings
> Are not thine own so proper as to waste
> Thyself upon thy virtues, they on thee.

My brother had secretly spent hours writing dozens of Shakespeare quotes all over his bedroom walls. They were some of the Bard's most famous lines, but to my mother's tired eyes, they were nonsense, gibberish, mumbo jumbo that defaced her home. She was furious.

She came home the next day with a can of paint and, ignoring my brother's pleas, covered the Bard's greatness with a sloppy coat of royal blue.

A few weeks later, after dropping Chiwetel off at school, she heard a tap on her car door. She leaned across the passenger seat and wound down the window. It was one of Chiwetel's teachers. She forced a polite smile.

"Is everything all right?" she asked.

"Fine. Actually, that's why I wanted to speak with you—"

She could barely hear him. The car radio was screaming, just a little louder than my little sister in the back seat, but she could sense it was serious. She snapped off the radio and tried to shhh her crying daughter so she could hear what the man had to say.

"As I'm sure you're aware, Chiwetel delivered a brilliant performance on the opening night of our new play."

She let out the breath she'd been holding. *Ha! He wants to talk about a play. A play.* She almost laughed as she tried to soothe Kandi with a toy, relieved that for once it wasn't bad news.

"Your son is gifted. Really gifted. I've taught drama for many years and I haven't seen anything like it."

My mother was clearly only half listening, but the teacher continued.

"I understand things are busy at home, but I would encourage you to watch him perform. We have three more shows, including one tomorrow."

"There's not much time for that, I'm afraid."

"The performances are held in that building at the end of the main drive, to the right." He pointed to the theater behind him. "It'd be a shame not to see you there."

"I'm sorry, sir. I work long hours."

"I didn't want to say this, but your son was the only child in the performance without a family member on hand last night. I really hope you can . . ."

My mother paused. The last thing she had time for was a play, on top of everything else. But for Nigerians, shame is a powerful motivator. She couldn't let this teacher think she was a bad mother.

She sighed. "I'll be there."

✍

The following night, she tiptoed into the school theater half an hour late, the room already wrapped in dark silence. She scanned the crowded hall before finding an open seat near the back left corner. She shimmied toward the orphaned seat past attentive parents, apologizing profusely for her mess of jingling keys, crunching plastic bags, and bulky coat. Fortunately, her neighbors were far too focused on what was happening on stage to care much. Once seated, she turned and looked up.

A few people moved about the stage saying something she couldn't quite make out. Then her son strode onto the stage, dressed in an intricate costume with bright colors, and started speaking. She didn't understand what he was saying, but he was speaking with a poise and confidence she barely recognized. A girl seemed to be pleading with him to spare her brother's life. She was asking for pity. Chiwetel boomed:

I show it most of all when I show justice;
For then I pity those I do not know,
Which a dismiss'd offense would after gall,

And do him right that, answering one foul wrong,

Lives not to act another. Be satisfied;

Your brother dies to-morrow; be content.

My mother knew very little about Shakespeare. She'd never read any of his plays and had very little understanding of the performing arts in general. But she could tell from the reaction of the white people around her that her son was doing well. She liked that a beam of light followed him around the stage as he interacted with the other actors.

When the curtains eventually fell and the spotlight went dark, the packed auditorium paused for a silent moment and then erupted in applause. Obiajulu stood when they stood.

The actors soon began their curtain calls, bowing to their cheering audience one at a time or in small groups, prompting new waves of applause each time. When Chiwetel's turn came, the roar was deafening. My mother's heart stuttered, and for a moment, she lost her balance. Steadying herself on the back of the chair in front of her, she surveyed the crowd. All around, the other parents and children and teachers were showering her son with love and respect. It was genuine. All this for her thirteen-year-old son and for nothing more than acting in a play.

For Nigerians, there is nothing more important than representing our families and communities well, no matter the obstacles. Chiwetel had in that moment achieved one of the single greatest accomplishments any Nigerian child can achieve. In an all-white auditorium in South London, four thousand miles from Nigeria, he'd brought pride to the village of Oghe.

If my mother could have frozen that scene, put it in her breast pocket and kept it close to her heart forever, she would have.

After the cheering stopped and people began to file out, she wandered down toward the front of the stage, not quite sure what to do next. From behind her, she heard a man's voice ask if she had enjoyed the performance. She turned and saw Chiwetel's teacher. His smile said everything.

"I'm glad you came."

"So am I."

She shook her head in wonder, still trying to comprehend what had just happened.

"Did you do this?"

"No, your son did. He is incredibly talented. I do believe this is the beginning of something extraordinary for all of us."

She followed the teacher out into the hallway, where a handful of other parents were waiting for the student actors. Chiwetel soon emerged and was bombarded by praise from the small gaggle. My mother, smiling quietly behind the throng, waited patiently for her turn to congratulate her son. It took a few minutes for her son's eyes to meet hers. He shuffled through the small crowd and stopped in front of his mother. She brushed a few specks of dust off his shoulders.

"Where on earth did you learn to do that?"

He responded with a bashful smile.

"That was a beautiful performance."

She leaned over and rubbed his back. "Your dad would have been proud."

They walked out into the night side by side, basking in the golden hue from the aging lampposts. Chiwetel ran around to the back of the car to dump his bags in the trunk, and in that moment alone, my mother allowed herself to feel the emotions that stirred deep inside. She was proud of what her son had done, but more than that, she was aware that he was channeling Arinze on that stage, that the passions she had sometimes discouraged in her husband had been reborn in his son.

By the time Chiwetel jumped into the passenger seat, she had wiped away her tears.

✍

My mother snuck off to the library the next evening on the way home from the pharmacy. Once the children were fed, and I was playing with Kandi on the cluttered living room floor, she retreated to her bedroom and pulled her new book from her purse.

Measure for Measure was the first Shakespeare play she'd ever tried to read. She moved through the prose slowly, by necessity and by design. She wanted to savor each sentence, each line, as she relived her son's performance from the night before. She could feel her husband in the lines, and she knew then, as she had the night before, that it was now her responsibility to develop her son's talent to honor Arinze in any way she could.

After that first performance, the school drama department began to offer my brother the lead role in *everything*. It didn't matter that he had to skip rehearsals at times to walk me home from school or to buy groceries for the house. They planned rehearsals to accommodate him when they had to. After only a few performances, he joined the Dulwich Youth Theatre, an acting club for young people that put on productions over the school year.

My mother continued her weekend trips to the library in that time. Whatever play Chiwetel happened to be rehearsing for—*The Tempest*, *Hamlet*, *Othello*—she'd borrow the book to study and practice on her own, usually while at the pharmacy. As complex as the language was, she wanted to know the plays well enough to follow along during his performances, to understand what was happening, what the characters were saying, so she could push him to be better.

Sometimes my mother would rehearse alongside Chiwetel, reading in her thick Nigerian accent Calpurnia's part to his Julius Caesar or Trinculo's to his Caliban.

Of all the things she thought she'd be doing as a Nigerian immigrant in South London, reading Shakespeare was not one of them. As she explored Shakespeare's world through the pages of his most famous works, she sometimes felt like a new immigrant all over again. But in this world, she knew she had a powerful ally.

Some weekends, she'd take Chiwetel to his drama club, dragging Kandi, who was still a toddler. My mother watched the rehearsals with rapt attention, often following along with her own copy of the play. They were a team, Chiwetel and my mom, each one pushing the other to be better.

Chiwetel began spending more and more time marching around the house studying his lines, his nascent baritone voice booming through our cold hallways. Most of the time, we had no idea what he was saying. He didn't care. He'd found someone who understood him at last, and he cherished every word.

My mother and brother were bonding over plays, but still clashing over the writing on his walls. I remember her shrieks when she walked into Chiwetel's room again one night to find the walls marked with even more Shakespeare. Although she now had a better appreciation for what he was writing, she still drew the line.

"Does it have to be all over every wall? I'll give you that one corner."

"But you just don't understand."

𝒦

As our lives got busier, my mother shared—and then enforced—a valuable lesson about time, to maintain Chiwetel's momentum in the performing arts and our focus on our studies.

Nigerians view time as a prized possession, a priceless and finite resource that must be respected and consumed wisely. My mother was always using her time for something productive: working, cooking, haggling, teaching. And as we got older, she taught us a simple but effective formula to ensure no time was wasted. She encouraged us to split our days into three equal parts: eight hours for sleeping, eight hours for school or work, and the last eight hours for working toward our dreams. Long before I reached my teenage years, the eight-hour rule was gospel.

The rule gave my siblings and me a head start in whatever we wanted out of life. It meant that Chiwetel spent countless hours practicing lines from plays *before* he scored an audition for a particular role, not afterward.

And that discipline is why nearly twenty-five years after discovering Shakespeare in that seventh-grade English class, my brother became the first African nominated for an Academy Award for the coveted Best Actor in a Leading Role category. His portrayal of Solomon Northup, a kidnapped Black man who never stopped fighting his enslavement in *12 Years a Slave*, was a transformational role for him, my family, my village, and millions of Nigerians who had never seen one of their own honored on the entertainment industry's biggest stage.

But I'll always remember the image of my big brother at family dinners, a script in one hand and a fork in the other, munching on his jollof rice while mouthing his Shakespeare lines.

Acting had brought him back to life after the car accident, and he couldn't tear himself away from his newfound passion, even to eat. I didn't appreciate what I was seeing at the time. Chiwetel was teaching me that when talent and passion merged with my mother's support and calculated work ethic, there was no limit to what we could achieve.

Chapter 6

My mother had lived in England for eleven months before she saw a washing machine up close. She'd heard rumors that life in London was so effortless that people actually relied on *machines* to wash their clothes. At the time, she was just happy to have a sink with running water to clean her limited wardrobe when she first got to Europe, so seeing an actual machine in person left her in awe.

The machines were uglier and louder than she'd imagined. Square frames, lots of confusing knobs and buttons. The way they shook and gyrated reminded her of a small airplane about to take off.

My mother first became acquainted with machines that could do such work in the early '70s at her first job in England, where she was tasked with washing bedsheets at a London hospital. It was a small facility, grim, dark, and crowded. Too many patients, not enough beds or doctors. The hallways were often littered with used gurneys and wheelchairs, and the small waiting area was always crowded.

Her job was to take the soiled linens, gowns, and towels to a large room in the basement with eight bulky washing machines. She was the only Black woman in the laundry department. She was also the youngest. The others were mostly white, English, and much older. They spoke with thick London accents she had to squint to understand, as though squinting somehow improved her hearing. Most times, no matter how hard she concentrated, she could never quite make it out. She'd smile and nod to buy time as she played back the conversation in her head, trying to crack the code.

In Nigeria, washing clothes was not complicated. She had three main outfits, and she was almost always wearing one of them. Washing clothes in England was something else. There was a whole host of decisions to make—settings, temperatures, and materials to be carefully considered. Sometimes she'd wash the linens or gowns more than once, just to marvel at the sheer satisfaction of having a machine do something that working people for millennia had been doing themselves. She'd laugh to herself in amazement: *What next, a machine to fold their clothes, too?*

Obiajulu's bright-eyed wonder came to a sudden halt after only a month, when she was transferred to ironing. The manager made it sound like a promotion, but she knew better. The irons didn't press the sheets for you. Disappointed, she grabbed her handbag and bagged lunch and reluctantly headed to the new world next door.

Rarely do we appreciate the small moments that change our lives for the better as they're happening. And so it was for my mother, who thought she'd been banished to a steamy dungeon when, in fact, the trajectory of her life—and mine—would improve exponentially as a direct result of this unwanted change.

The job itself was mind-numbingly boring and mundane. For seven hours each day, she pressed heat onto linen. Then she'd fold the crisp sheets into neat rectangles, placing them into a cupboard for someone else to take upstairs.

At first, she really enjoyed watching the wrinkles evaporate. But the lack of human interaction got to her. There were other women in the room, but the simple work required no real coordination with others, and she was usually too shy and self-conscious to make small talk with the English people on her team, so she was alone with her work most of the time. The monotony made the days blend together, and she probably would have asked for another transfer if Mary hadn't arrived.

No one had mentioned anything about a new addition to the ironing crew, but here she was, a tall, somewhat corpulent Black woman whose hips swayed with purpose as she moved about the room. There was something about her gait that said her life had texture. And when my mother heard Mary speak for the first time, she noticed the unmistakable melody of a thick African accent. *Another Nigerian!*

My mother had lived in London for a year, and she had yet to make any real friends. There wasn't much time for that sort of thing amid her sudden pregnancy with Obinze, her long days at the hospital, and the schoolwork she'd take home each night. She'd tried being friendly with a few British people but found herself too self-conscious about her village ways to genuinely connect with them.

In Nigeria, friends were much easier to come by. Obiajulu had enough siblings to fill a soccer team; between her sisters, her cousins, their friends, and their cousins, she was surrounded by an army of companions at all times without ever having to leave her street. But one year after moving to England she still didn't know the names of her next-door neighbors. She looked at the ironing department's newest hire and was hopeful that her friendless existence in this cold new country might be coming to an end.

My mother noticed right away that Mary was a fast walker. That might not mean much to most people but it certainly made her stand out from all the other Nigerian women my mother knew. They mostly tended to walk slowly, no matter the circumstances, as though content to soak up every one of life's little offerings—the good and the bad. But Mary, she was a woman on a mission.

They didn't speak on her first day at work. On the morning of her second, my mother felt a tap on her shoulder.

"Who can we get to fix number four?" Mary's accent alone made Obiajulu feel better.

"Huh?"

"The iron at station four. It's not working."

"It's happened to me before. You just have to unplug it and start over. I'll help."

Mary was slightly older than my mother, and much more polished. She walked and talked with the confidence of a woman who belonged. Once the irons were straightened out, they shifted the conversation to life beyond work. Mary was from Owerri, a three-hour drive from my mother's hometown in Enugu. They shared painful stories about the friends and family they'd lost during the war, the countless Igbos with missing limbs who roamed the streets looking for work. The male population had been decimated, so women had become the primary bread-winners and, in many cases, the last line of defense in keeping Igbo families from falling apart, although many did. And too many widows begged for money outside churches and markets, recognizable by their freshly shaved heads and white gowns.

Despite the devastation, Mary and Obiajulu were sure that their small towns and big cities would be rebuilt someday. They, and other recent immigrants like them, cared too much to let their people down.

The women strolled through the basement hallway in matching cleaning uniforms as they spoke, laundry baskets in hand. My mother was so engrossed in the conversation, she didn't realize her shift had already ended. She looked at her watch and panicked, apologizing profusely as she ran toward the main exit and to her husband waiting at home. On the bus back to their flat in Brixton, she felt a flicker of belonging for the first time in London.

As they got to know each other better over the coming weeks, the conversation moved away from the past and toward the future. Mary was ambitious. She wanted to start her own business, perhaps even a real estate agency. Most Nigerians who were doctors or lawyers back home knew that coming to England meant starting over, driving taxis or cleaning offices. Not Mary. Even in her cheap hospital uniform, she was already planning the rich ways her life in London would unfold—and how she would use her success to help her family back in the village.

The first eighteen years of Obiajulu's life had been defined by struggle and strife in a nation torn apart by political and ethnic division and a brutal civil war. She'd experienced the kind of poverty that most people only read about in newspapers. There weren't enough jobs to go around after the war. So, her father moved his children to a tiny bungalow on the edge of a garbage dump where she and her six siblings slept, bathed, and cooked in the same room. Later, he was forced to sell the family's meager possessions, including his clothes, to pay for Obiajulu's last year of high school. He openly discussed marrying her off to any Nigerian who made a decent living abroad; anything to have one less mouth to feed.

Now that she was in London, she didn't want to struggle anymore. But she had never quite imagined the kind of ambition and material success—for herself, at least—that Mary was describing.

The women spent almost every day together. They ate lunches of home-cooked moin moin as they plotted their futures. Mary did most of the talking and my mother listened with rapt attention.

After long days of hopeful conversation, my mother would lay awake for hours at night thinking about the possibilities. She knew she'd be a doctor's wife someday, and that was something, but maybe she could be more. Those days in the early '70s marked the beginning of my mother's belief that she and her family could be, do, and achieve whatever they wanted in London.

It didn't matter that she had lived with pain and hardship in her homeland for as long as she could remember. It didn't matter that Britain was in recession and most people were simply grateful to have any job at all. It didn't even matter that she was an African immigrant living and working in a system that worked against immigrants. Mary opened my mother's eyes to the possibility of having a real career, owning her own business, and creating a future she controlled. Mary was living in a world with no ceilings. Maybe Obiajulu could, too.

After cleaning the kitchen one night, my mother shuffled into the small living room, where my dad was studying at the table. It was time to share her ambitions out loud. She gently sat down and starting telling Dad about Mary, about her plans to start her own business, to own something, to work for herself.

She talked about her interest in the sciences. Although she didn't want to become a doctor like him, the war had taught her the importance of medicine. Even after the war, Nigerians in the east were still dying from malaria, cholera, and other waterborne diseases. Maybe she could pursue pharmacy. Maybe she could own her own store, make her own hours, and contribute to her community back home. Being a pharmacist would be respectable in the eyes of the village, and Mary had talked about pharmacies being recession-proof.

"Who is this Mary? She has woken something inside you all right. I like what I'm hearing."

Within a few weeks, my dad was convinced my mother was serious. He soon was as excited as she was. She eventually enrolled in the University of London to pursue her pharmacy degree, and during her studies, they began to map out a plan to open a pharmacy of their own. They wanted something in a residential area that would generate a loyal customer base. And, of course, it had to be affordable. My father was still in medical school, and my mother's part-time work at the hospital barely paid for groceries.

She'd never set foot in a bank in her life, but a year after graduation, my mother needed a loan to get the business started. She wore the most professional-looking dress she had under a button-down coat. My dad wore a white shirt and black tie. They had practiced their lines countless times in front of the bathroom mirror, but standing outside the bank, my mother suddenly couldn't recall what she was supposed to say. *I guess it's in God's hands.*

She took a deep breath as she and my father pushed open the bank's glass door and scanned the room, looking for the manager's office. He welcomed them coolly and closed the door. My father opened the pitch as the bank manager scanned their documents silently.

"So, you'll run the pharmacy?" he asked my mother.

"Yes, sir. I'm fully qualified. And we've picked out the perfect location—right off a high street, which is good for accessibility," she said, stealing a glance at my dad to make sure she was doing okay.

An hour later, my parents walked out of the bank, their heads down. They walked to the end of the block and checked to make sure no one was looking. Then they erupted, screaming, jumping, dancing, and hugging. They'd got the loan.

A few months later, they held the keys to an empty retail space just off Brixton Road. After a minor renovation, some new shelving, and freshly framed degrees on the wall, they opened their own business. It was real. The village girl from Oghe who studied long division with sticks in the dirt as famine killed her neighbors was now a legitimate business owner in London.

Decades later, my mother would recall how the pharmacy really saved us after Dad passed away. That cramped store in Brixton with just two aisles, an unlikely symbol of one immigrant's ambition, gave her the flexibility and the resources to raise four children as a single mother.

Thank God for Mary.

~

This idea of *uplifters*—people of similar backgrounds who can open our minds to new possibilities—is not unique to Nigerian culture, but it was a defining principle in how my mother raised us. The book club, the personal study time, and the discipline were essential. But we needed more, especially as we began to confront the inevitable challenges people of color face in overwhelmingly white societies. We were consistently among the only Black or Brown children in our classes and despite our academic success, each of us had trouble fitting in at times.

Chiwetel, a thirteen-year-old Nigerian boy who scribbled Shakespeare lines on his bedroom walls for fun, often struggled to find acceptance among his peers, even after his acting success lifted his status at school considerably. Older boys sometimes poked fun at his name and made cruel comments. He would pretend not to notice, but of course he did. Obinze was once locked in a closet for hours after school as the other children hurled racist abuse at him beyond the door. My parents removed him from that school shortly afterward.

The sense of not belonging also followed me into a silly game we played almost every day at school called Kiss Chase. The rules of the game were as simple as the name suggests. The boys and girls would stand on opposite sides of the concrete playground until someone shouted, "Go!" Then the boys would chase the girls and try to kiss them. I remember being so excited the first time I played. There I was, a seven-year-old running and shrieking with delight. Except, when I finally turned to see who was chasing me, no one was there. I remember thinking that maybe I was just a really fast runner and the boys couldn't keep up. But then it happened again. And again.

I grew to understand that the white boys simply weren't interested in chasing the Black girl. I was ashamed and embarrassed, as though something were wrong with me. The worst part, as I think back now, was that I continued to play the game day after day anyway. I knew that no one would come after me, that I was not wanted, but I ran and screamed and pretended to laugh just the same.

My mother had prepared us to excel in academics, but she did not have an answer for this. Despite all our accolades, achievements, and praise, school was not a place where we ever felt safe. And my mother had no idea how to make it better. She'd spent her entire childhood in Africa surrounded by Africans. She'd survived civil war and could teach us lots of things about perseverance and hardship, but if you asked her how to deal with racism, she was stumped. Now, for the first time in her life, she was beginning to understand that being Black brought with it a host of other challenges that no amount of A grades could compensate for.

My mother was angry for us and she was afraid. She knew in those moments, and probably long before, that hugs and words of encouragement would not solve our problem—not in a country dominated by both class and racial divisions for decades. She wanted with all her heart to make it better.

She spent time searching her own life story, her own struggles and triumphs, for any experience that might help. She thought about Mary in the ironing room at the hospital all those years ago and realized how much her spirit had been lifted by someone who looked like her, sounded like her, and had a strong sense of purpose. Mary, directly or indirectly, changed her perception of herself as an African immigrant, encouraging her to break through boundaries that seemed unbreakable.

That was what we needed. We needed our own Marys. People who looked like us, with equal ambition, who faced down similar racial challenges and thrived anyway.

My mother immediately set about in search of these uplifters, which she referred to in our language as *ndi eji amatu*, which loosely means "those who set the standard." After I came home from school one day, she burst into my room and handed me a page torn from a newspaper. She was beaming with pride as she watched me read it. The article was about the musical duo Lighthouse Family, a popular British act that was topping the UK charts at the time. She pointed at the name of one of the members, Tunde Baiyewu. I asked if he was someone we knew.

"No," she responded, "but he's a Nigerian. And that's what's important."

She began to stack our living room bookshelf with the works of artists, novelists, and playwrights from West Africa, introducing us to the folktales of Cyprian Ekwensi and the post-modern fiction of Ben Okri.

She continued to seek out newspaper clippings about other Black people, some born and raised in South London like us, who had gone on to excel in their fields, anyone who had reshaped the world's perception of what it meant to be Black. On days when we weren't doing book reviews at the dinner table she'd often read their accomplishments out loud and force us to discuss them. She drilled it into our heads that the people in the stories were just like us, and if we worked hard enough, we could have what they had.

Years before Chiwetel was accepted into one of the country's top drama schools, she found articles about Black students who'd studied there, and stories about awards they won. She also found references to Black British musicians—people like the composer and conductor

Samuel Coleridge-Taylor, the pianist Winifred Atwell, and the singer-songwriter Joan Armatrading. She made a point of seeking out world-famous Nigerians, too—like the soul singer Sade and the R&B star Seal.

And it was more than just dinner table conversation. Shortly after I turned twelve, my mother began to plaster my bedroom walls with posters and articles of African success in Europe and beyond. I came home from school one day to find my bedroom mirror missing and no fewer than five newspaper clippings crudely stuck to the walls. In each, taken from a collection of *The Times*, *The Guardian*, and *The Telegraph*, there were references to people who looked like us who had achieved something extraordinary. I remember first learning about the Nigerian playwright and novelist Wole Soyinka's life story in detail—from his imprisonment during the Biafra war to his Nobel Prize in Literature two decades later—and studying up on Paul Boateng, one of the few prominent Black British politicians at the time.

Their stories may have been inspiring, but as a girl on the verge of becoming a teenager, I desperately wanted my mirror back. When I confronted my mother about it, she said simply, "Less focus on how you look, more focus on what you can become."

I never saw that mirror again.

I was maybe six years old when my mother met a woman in London, Mrs. Chinegwundoh, who was actually from our home state, Enugu. She'd left Nigeria just before the war and spent more than a decade in various teaching positions across several London schools. She became an important resource to help my mother navigate the school system for my sister and me. But she was more than that.

Just before I turned thirteen, my mother burst into my room at 6 a.m. one morning to share the big news. It was about Mrs. Chinegwundoh's son.

"Can you believe it? Mrs. Chinegwundoh's son has just graduated from Cambridge."

"Mrs. Chine-who?"

"Gwundoh. Her son."

I stared blankly.

"Come on. You met them last year. She gave you that toy chemistry set for your birthday."

She sat on the edge of my bed, talking loudly and gesticulating eagerly, marveling at his achievement. I pulled the covers over my face and tried to force myself back to sleep.

The truth was that neither of us knew the Chinegwundoh family very well. But my mother was acting as though I had graduated from the elite university. In the coming weeks and months, she'd give us regular updates about John Chinegwundoh, the young Cambridge graduate. We learned what he studied, his interests, his plans after graduation. I had met John in person only once or twice before then, but it felt as though we knew his entire life story. I'm sure he had no idea, but this young man became as popular in our home as a famous singer or actor.

My mother began looking for opportunities to get us in the same room. It wasn't enough to talk about him; she needed me to see the reality of his success and work ethic with my own eyes. We went to Mrs. Chinegwundoh's small terraced house in Brixton one Sunday afternoon under the auspices of hearing about her recent trip to Nigeria. As the women lamented in the kitchen about the current state of affairs in their home country, my mother not-so-subtly gave me a look. John, who was home for the holidays, had just walked into the living room. He was several years older, and I'm sure he had better things to do than humor a thirteen-year-old, but I knew what my mission was. Scanning the textbooks stacked on the table next to him, I saw my opening.

"Were those for your classes at Cambridge?"

He nodded. "And so were those and those," he said, pointing at several other piles of books around the room.

He told me he'd attended Cambridge's Emmanuel College, which societies and clubs he'd belonged to, and how one day he hoped to make a difference in the world of medicine.

We spent just under an hour like that, a university graduate sharing details about his life and studies with a thirteen-year-old he barely knew. The conversation was awkward at times, but it worked. I left the Chinegwundohs' that day knowing that someone just like me could go to one of the best universities in the world and graduate top of the class.

❧

As was the case with much of what my mother did for us, her focus on uplifters helped her, too. She faced serious personal challenges long after the shock and pain of my father's death began to fade. She was constantly worried about us, worried she was failing at the insane juggling act her life had become. And she faced the real prospect of violence at the pharmacy, more often than we realized.

Years later, she would tell me about the night she stayed late to re-stock the shelves—cotton buds, Chupa Chups lollipops, sunscreen: the odds and ends that ensured she had regular customers. As was almost always the case, she was in the store alone. By the time she had stacked the cardboard boxes in the back room and switched off the lights, it was past ten o'clock. Her car was parked on the street just thirty or forty feet from the back door. The neighborhood was flooded with streetlights, so even when the streets were deserted, as they were that night, she rarely worried for her safety. But almost immediately after she locked the back door and turned toward her aging Honda Civic, she heard the shuffling footsteps of someone behind her.

She spun around and was immediately confronted by a young man holding a sharp blade, his eyes peeking out from under the rim of a dark baseball cap. He moved toward her and, after scanning her up and down, gestured with the knife to her rings.

She looked down at her left hand, at the simple engagement ring and wedding band that had brightened her fourth finger for more than two decades—and every day since Arinze's death.

She looked at the knife again and back at the rings, glittering symbols of the commitment she still felt for the love of her life. She didn't want to let them go. The young man stepped closer, too close, and she surrendered. She tugged at the rings, shaking, pulled them off her finger, and held Arinze's treasure in her open hand for the boy to take. He did, and scrambled off into the night, leaving my mother to cry alone.

Her eyes were red and wet when she came home that night, but she wouldn't tell us about this encounter, or the many other robberies at her store, until years later. She was still as focused as ever on seeking out people in our community who were doing well, driven by a new sense of urgency to make sure that I could someday escape that neighborhood and the dangers that she could not.

*

My mother had a white binder filled with pages of plastic sleeves she loved to read in bed. She filled the sleeves with her favorite collection of the clippings she had plastered around the house to expand our ambitions—story after story of Black people who overcame and soared. She loved to add to her collection, reorganize the pages, and reread her personal favorites. Sidney Poitier's story was particularly appealing given Chiwetel's budding acting career. Margaret Busby, a Ghanaian who was Britain's first Black female book publisher, was another special one.

Whenever we complained about a difficult school project, a test, or a bully, she would take out the binder. Invariably, she would leaf through the scratched plastic pages and find just the right story to quash our doubts.

"Have I told you about Jocelyn Barrow?"

"Yes, Mum."

"She played a major role in the fight against racial discrimination here in the sixties, then became a governor of the BBC—the first Black woman to do it. I don't want to hear any more complaints about how hard your homework is. Go study."

It was an effective strategy, even if it was annoying at times. She knew every story in her binder by heart, and eventually so did we. I became a walking encyclopedia of Black British success before I became a teenager, keen to emulate the very people I'd heard about from those newspaper clippings my mother kept under her bed.

Sometimes she would force the binder upon us when we were least expecting it. More than once, she sprang into Chiwetel's room, binder open, while he was in the middle of rehearsing a monologue.

"Have you seen what Colin Salmon is doing next? He's in the new James Bond movie. They're filming soon. Read this."

"You've shown me that story five times already."

She would also pounce on Obinze. He was studying biochemistry and physiology at university by then but constantly talked about starting his own business one day. Mohammed Indimi was among those featured in the clippings of African entrepreneurs she'd brought back from one of her visits to Nigeria. She forced it upon him as soon as he crossed the front threshold.

Before his jacket was off, she'd read aloud some of the highlights: "This man started out selling animal skins in his village. Now he runs an empire. One of the most successful men in the country. Look."

But these uplifters were more than just fluffy inspiration. I remember the first time my mother called me downstairs to catch a glimpse of Femi Oke, a Nigerian British TV host presenting an educational program on the BBC. Oke was always so poised, so charismatic during her reports.

"You see that?" my mother would ask. "You see what we can do?"

I usually rolled my eyes whenever she forced us to watch one of her African success stories on television, but there was something different about Oke.

Seeing her, and by extension my culture, displayed so visibly on national television blessed me with a sense of belonging. I felt seen and heard in a way I hadn't before. If Femi Oke, a Nigerian woman, could make it this far, then perhaps I could, too.

Several years later, after I'd moved to America to study journalism, I switched on the television one afternoon and saw Oke reporting the news on CNN. I felt so proud of how far she'd come, so inspired, that I immediately sent her an email, telling her what it meant to me, what it meant to my family, to watch her on television all those years earlier.

I didn't really expect to hear back. She was a big star who spoke to millions on television every day, and I was a random student she didn't know. But she responded within minutes and gave me her phone number. We spoke for more than an hour that night, and I listened as she gave me invaluable advice about making it in the newsroom.

Years later, when I applied to CNN, I reached out to her for guidance. Oke had since left television to work in radio but she guided me through the interview process and gave me tips for the screen test. Her generosity was incredible. And when my big opportunity came, it made all the difference.

Femi Oke was a Nigerian woman who had been helping me, directly and indirectly, for almost a decade. Seeing her on screen all those years ago was not only a beautiful nod to my heritage, it actually changed the course of my entire life.

The uplifters had done their job.

It's a strategy I'm beginning to implement with my own children by giving them examples of uplifters in the country where they live and beyond. I don't yet have a white binder with plastic pages like my mother's, but I have lots of material to choose from when I do.

I recently discovered that Poland's first Black MP was Nigerian, as was Italy's first Black senator. The first Black female president of the *Harvard Law Review* is also Nigerian, as is the first woman to run the World Trade Organization.

In 2012, the British media dubbed a Nigerian family, the Imafidons, one of the smartest families in the country, after two of their children started college by the age of thirteen. In 2021, a ten-year-old refugee who was once homeless became America's newest chess master. He too is Nigerian.

The sports world is no different. As of 2021, both reigning UFC middleweight and welterweight champions were Nigerian. The NBA's most valuable player in 2019 and 2020 was Nigerian. Nneka and Chiney Ogwumike recently made history by becoming the first sisters drafted first overall in the WNBA. Yes, both are Nigerian.

Such successes were harder to find when I was growing up, but my mother found them nonetheless. The binder, the wall clippings, the people on TV—they were constant reminders that we were not bound by our neighborhood, the pain of our loss, or the low expectations of the world around us. It was just the opposite. We were expected to thrive at all costs. And so we did.

Chapter 7

There's no easy time to fall in love. The hopeful longing, the anxious desperation, and the sleepless nights can be beautifully brutal even in the best of times. But that summer of 1965 was not the best of times—not for my teenage parents, and not for a country on the edge of war.

At just sixteen and fourteen, Arinze and Obiajulu were too infatuated with each other, and perhaps too young, to understand the dangers that lurked all around as Nigeria's many tribes and factions jockeyed for power in those early years after the nation's independence.

My parents had their own dangerous missions to attend to. After their chance meeting in that restaurant and the magic that followed in the village square, my parents wanted nothing more from life than to be close to each other. But in a village like Oghe, boys and girls couldn't show any signs of public affection until after they were properly married. If they did, rumors would start and reputations could be ruined. And once ruined, my grandmother warned, reputations in a small community like theirs would be difficult, if not impossible, to rebuild.

Almost every morning, Obiajulu would race her bike down the blood-red dirt path on the edge of the forest to Arinze's modest home, hiding behind a tree until his parents left for the day. Other days, they would spend hours whispering in a clearing in the bush, protected by the umbrella acacias and thick shrubs. Many afternoons, she would march down to the village restaurant with girlfriends, gather around a table outside, and feign interest in their conversations while passing a note to the handsome boy who served their lunch.

The unwritten laws of their village prevented them from diving headfirst into the kind of unrestrained summer love that you may have known in your life, but their feelings were no less real. And aside from Evelyn, the cousin who introduced them, no one knew—not any of my mother's six siblings or my father's three, not the village igwe or the parish priest or the medicine man who guarded the shrine, not the aunties who sold cocoyam in the market or the men who harvested it. Obiajulu and Arinze were in love. But it was their secret. And after only seven weeks, it nearly came crashing to an end.

By late summer, it was time for Obiajulu to return to school near Jos, a city almost a full day's train journey to the north. For two teenagers, it was a world away. They were devastated. Obiajulu desperately wanted to stay behind—she dreamed of leaving school so they could run away together. But in the end, as Nigerian children usually do, she did what was expected of her. When the day came, my mother loaded her suitcase into the back of my grandfather's Peugeot 404 and blamed her tears on the dust swirling in the hot summer wind. It was too risky for Arinze to see her off in person, but as my mother hugged her cousin goodbye, Evelyn slipped a note into her hand, a note from my dad she still has today. Just three words: *I love you.*

Nigeria celebrated its independence from Britain on October 1, 1960. By October of 1965, two months after my mother returned to the north, the earliest bloodshed had begun. It was a horrific, if predictable, result of the power vacuum created when the colonial power left a country of 50 million people teeming with three hundred different ethnic groups, each with their own priorities, values, and conflicting beliefs.

Even before the violence started, Nigeria was always a hard country to hold together.

Resentment simmered among each of the main tribes as far back as the early 1900s, but there were two tribes who never quite sorted out their differences: the Hausa-Fulanis in the north and the Igbos in the east. When the Hausa-Fulani northerners gained political power after the 1964 elections, the country's first after independence, Igbos claimed election fraud. When the northerners refused to compromise, the Igbos decided the time for talk was over.

In January 1966, a handful of officers, most of them Igbo, ordered their soldiers to carry out a coup. Several northern politicians and military leaders were murdered. In retaliation, the northerners then began a systematic ethnic cleansing of Igbos that would later pit neighbor against neighbor and armed soldiers against women and children. The northerners killed any and every Igbo they happened to lay their eyes on, hunting them down in churches, schools, and markets. This wasn't hard to do. The Igbos were a minority in the north and were easily distinguishable from the rest of the population by their Westernized clothing, last names, and separate language.

The explosion of violence that followed became known as the *pogroms*, a chillingly sterile term to describe a systematic massacre that left tens of thousands of Igbos dead—many disemboweled or beheaded by machete—over the span of just a few months. It may be the most savage period of ethnic cleansing in the modern era that most Americans have never heard of. At the height of the bloodshed and the civil war that followed, Western nations were focused on their own headlines—the assassination of Martin Luther King Jr., the Arab-Israeli Six-Day War, and man's first landing on the moon—all of which drowned out the cries of thousands of Igbos being slaughtered by their fellow countrymen from 1967 to 1970.

There is some dispute over how many Igbos living in the north were ultimately killed—the higher estimates are close to 60,000—but there is no dispute about the mass exodus triggered by the violence. More than a million Igbos fled east in 1966, literally running for their lives as their northern neighbors joined northern soldiers in hunting down and butchering any and all Igbo men, women, and children who couldn't leave fast enough. Machete was often the weapon of choice, but the military used guns, too. As Nigeria braced itself for a brutal civil war, the lucky Igbos became refugees, homeless and hungry.

My mother would become one of them.

ৼ

Separated by 350 miles of dirt roads, rivers, and forests, Arinze was in agony those first few weeks after Obiajulu left the village. Desperate perhaps just to say her name out loud, he asked his parents one afternoon what they knew of the young girl, Obiajulu, who spent summers at the Okafor residence. His father eyed him suspiciously.

"Why are you asking?"

"No reason. I met her through Evelyn. Thought you might know her."

His father folded his newspaper in half and set it aside.

"Aren't you supposed to be at the restaurant today?"

Arinze nodded his head, yes. "I'm on my break."

His father continued: "I don't know her, but I know her father, Ignatius. He works for ATMN, the tin mining offices in Rayfield, just outside Jos. Decent man. What do you want with that family?"

"Nothing."

But Arinze had what he wanted. By the end of that week, he'd packed a small cloth bag with two changes of clothes and a few books and told his parents he was heading back to school early. Many Nigerian schools were boarding schools in those days, a remnant of British colonization. But instead of heading to Uwani, he boarded a long-distance train heading north.

With all the stops, it took almost a full day by train and bus as he traveled through Nigeria's heartland: the lush green vegetation of Enugu, the wild rivers of Benue, and the sprawling savannas of Plateau, dreaming of the girl who'd put him in this trance. At sixteen and the youngest of four, Arinze was already an adventurer. He often traveled about on the back of trucks with people he hardly knew, but he'd never made a solo journey like this before.

He should have been exhausted by the time he stumbled off that bus in Rayfield. Instead, he was vibrating with excitement. Obiajulu was close. He could feel it. He only had to ask one person to find out where the local tin mining offices were. ATMN was a major employer in town, one of the many private companies still operated by the British after independence.

Arinze walked up a steep dirt path and paused to catch his breath as an imposing brick building came into view. He climbed up one flight of stairs to the main lobby and stated his name and business for the office messenger who quickly identified his target. He waited by the window, staring out at Rayfield's sprawling expanse of colonial bungalows. Rayfield was home to a mostly white enclave of former colonial employers who lived in relative seclusion. With access to sports clubs, swimming pools, cooks, and gardeners, the residents enjoyed the sort of lives that left most locals in awe. Arinze had heard rumors that people there played croquet and snooker and ate cheese, bacon, and other delicacies straight from Europe.

For a few minutes my teenage father stood there at the window, lost in fantasy, until the messenger tapped him on the shoulder and politely asked him to follow. Into the heart of the winding building they went, up another flight of stairs, before he knocked softly on an office door and pushed it open. Ignatius was sitting behind a sturdy desk. He looked at Arinze quizzically.

"Can I help you?"

"My name is Arinze, sir. I'm from Oyofo-Oghe."

"What is this about?"

"I believe you know my father, Michael."

Ignatius looked unsure.

"Michael. The carpenter. He does most of the handiwork at our church."

"Oh yes." Ignatius removed his glasses and folded the stems. "What are you doing in Rayfield? You're a long way from home."

Arinze had practiced this. It would work. It had to work.

"We're researching the tin mines for a school project. My father suggested I stop by to say hello."

Arinze handed him a letter of introduction he'd forged from his dad. Ignatius scanned the lines quickly. Arinze continued:

"We have to write a whole dissertation on the history of tin mining on the continent and how the British—"

"Where are you staying?"

"With classmates. In Bukuru."

"Nonsense. Stay with us. You're family. My wife would be delighted to see you. And I have two teenagers close to your age."

Arinze tried not to smile.

"Thank you for such generosity, sir."

Where the Children Take Us

At that very moment, Obiajulu was getting her hair braided in a small hut near the market. As the woman pulled and tugged her afro, she passed the time by thinking about Arinze. They'd been apart for less than a week. She wondered if he'd received her first letter, whether he missed her as badly as she missed him, and whether her heart could wait until the next summer to see him again. She squeezed her eyes shut and tried to remember his face, every line and curve. If she tried hard enough, she could hear his voice and feel the warmth of his presence. Just the thought of him made her heart rate spike. She lived mostly in her imagination that first week back in Rayfield. It was where she was happiest. Arinze was her waking thought, her source of joy, excitement, fear, and anguish throughout the day, every day, until she closed her eyes at night and dreamed about him some more.

She was still with him in the deep recesses of her mind when the braider tapped her on the shoulder and signaled she was done. Obiajulu mumbled a thank-you, handed over a few pennies, and shuffled out the door, wandering along the busy main road toward her parents' house in a happy daze. She reached the front door and turned the knob, hearing voices as she entered the room. Her parents seemed to be entertaining a guest.

She scanned the room and gasped as she saw the very face she'd just been dreaming about. It was Arinze's. He was here, in her house, sitting in the chair next to her father. Obiajulu was never a girl of many words, but in that moment, she couldn't have formulated a proper sentence for all the money in the world.

Arinze smiled, stood up, and approached her, stretching his hand toward hers.

"Nice to meet you."

My mother was wide-eyed and frozen to the spot. She cautiously extended her hand to touch his.

Ignatius interjected: "Obiajulu, this is Arinze, an old family friend."

She refused to blink, for fear he might disappear if she closed her eyes, even for an instant.

"He'll be staying with us for the next week for a school project. You can show him around town tomorrow afternoon."

She was desperate to touch him again, but kept her distance until her parents left the room. As soon as the coast was clear, she reached for his hands. He stepped back and shook his head.

"Soon, but not now."

"What are you doing here? How have you ended up in my house?"

He told her of his journey, that he couldn't bear another day, another moment, without being close to her. He'd traveled hundreds of miles by train and would have traveled thousands more if he had to.

This boy must be mad, she said to herself.

Over the coming days, Arinze and Obiajulu tried to keep up the charade as best they could. Obiajulu would sometimes pretend to be running errands and invite Arinze along; other times they'd leave the house separately and meet up in town. Her family had lived in Rayfield for as long as she could remember, so she knew it well. She showed him the tennis courts where the white people chased yellow balls, the markets where her mother bought goat meat and garri, and the brand-new postal agency where you could send a real-life telegram.

She tried to impress Arinze with the slice of sophistication that came from living part of the year outside the village. "I just love apples," she told him. "I tasted one once, and they're so sweet." It was almost impossible to find apples in Nigeria in those years. Arinze surprised her when he said he'd tasted them several times. Obiajulu considered herself the more cultured of the two, but she was starting to see things differently.

He responded with a power play of his own: "Did you know that in London oranges are actually orange?" She thought he was joking at first. In Nigeria, of course, oranges were green, and if they were really ripe, they eventually turned yellow.

It went on like that for almost a week. Quiet conversations on secret walks as they fought the temptation to touch hands in public. Obiajulu's mother, who sometimes spotted them in the market, grew suspicious. Their relationship lacked any of the awkwardness of a first-time encounter, and errands that would normally take my mother fifteen minutes were now taking two hours when she was with Arinze.

But Arinze and Obiajulu were oblivious to the world around them, not noticing my grandmother's menacing stare, nor the newspaper headlines and radio reports of political battles and flashes of violence. Nothing else mattered, especially as their time together sped toward a second ending. School was about to start for both of them, Arinze back in the east and Obiajulu nearby in Jos. They knew there was no avoiding it but the mere thought of having to say goodbye came with a lingering sense of dread.

After only five days together, Arinze headed back to the bus station for the daylong trek in the opposite direction from his heart's greatest desire. Obiajulu met him there, trying with all her might to keep her eyes dry. They told each other they'd only be apart for a few months. He would find a way to see her again around the Christmas holidays.

She buried her head in his chest, not caring who might see, and he wrapped his arms around her tiny frame. He kissed the top of her head and gently broke away, climbing the bus stairs with all the determination he could muster. He did not look back.

They did see each other again that Christmas, when Obiajulu visited family back in the village, and again the following Easter, but their relationship existed mostly on paper in those early months. They wrote to each other constantly, once a week or more, sharing life the only way they could. But my mother was alone in school in Jos the following May when the massacres began.

It started off as mere hearsay. Rumors that Igbos in other northern cities—Zaria, Sokoto, and Kano—had been attacked by angry northerners. Men and women, it didn't matter, shot, stabbed, or clubbed without a word, apparently in retaliation for a coup that none of them had taken part in or really understood. My mother's parents, Ignatius and Caroline, were worried but hoped the worst of the violence would come and go quickly.

So did most of Jos's 100,000 residents, who went on with life as normal at first. The vast majority didn't have any reason to worry. Igbos made up only about 10 percent of Jos's population, and there were no glaring signs of friction, at least not immediately. Ignatius kept going to the mining office in Rayfield each day while Caroline cared for their two youngest, just two and four at the time. Their four middle children attended local elementary schools, while Obiajulu, their eldest, studied at a boarding school a short drive away.

The human mind can be remarkably resistant to the darkest realities of war before it sees them, but things changed for Caroline the morning she bumped into a neighbor crying hysterically in the middle of the road. Through her sobs, the neighbor explained that two of her children had been killed while at school in Kaduna. A group of boys with machetes had stormed the school, targeting Igbo children, and during the stampede to escape, her children had been trampled on and crushed. Schools in some northern states were suddenly encouraging Igbo children to flee, unable to guarantee their safety.

Suddenly convinced, Caroline sprinted home and ordered a cousin staying with them to retrieve Obiajulu from school in Jos. She scooped up the toddlers and raced to collect her other four children. Along the way, she heard frantic talk of men with clubs roaming the streets. Someone said British missionaries were offering to hide Igbos in their attics and storage rooms, while white families in Jos were hiding Igbo children under their beds. There was smoke billowing in the distance.

Obiajulu rode home from school with her cousin in silence, upset that she was made to leave several weeks early. She didn't understand what the fuss was about. Who cared about a few isolated skirmishes? She was only halfway through her exams, exams that she had spent weeks preparing for and desperately wanted to finish.

But as they drove through the outskirts of Jos, she was struck by the eerie emptiness of the streets. Outdoor market stalls, normally swarming with people, were closed. School playgrounds sat still. Churches that offered daily services were padlocked shut. She scanned the streets seeking some explanation, but none came. Then, through the passenger window, she spotted the first of the horrors. Two, maybe three people—at least one seemed to be a child—lying lifeless and bloody in dusty heaps on the side of the road. There were no passersby stopping to help. The driver accelerated and told his passengers to stay low. As they sped toward Rayfield, she peeked out the window and saw more bodies, two women this time, their clothes soaked in red.

The car lurched to a stop in front of Obiajulu's house and she scurried inside to see her mother swiping everything off the mantelpiece—pictures, dishes, candles—into a woven bag. She screamed at Obiajulu to get the children ready.

"Ready for what?"

"We're leaving. Now. Right now."

"What's happening?"

"Whatever you can pack in five minutes. We need to go. Now!"

Ignatius stood in the middle of the room, pleading with his wife to reconsider. There was danger outside, sure, but no worse than the Jos riots of '45 or the Kano riots of '53. Ignatius believed they would be safe inside their own home. Not to mention that he had a job, a good job, and his white bosses would surely replace him if he ran off like this.

Caroline raised her voice. "People are dying! What's wrong with you?"

"You expect me to abandon everything we've built because of a few rumors?"

"Think about the children! Even the white people are running back to England. And you want to stay?"

It broke her heart, but Caroline realized there was room for only one compromise in that moment.

"If you want to stay, stay. We're headed back to Enugu. We'll wait for you there."

Caroline and Obiajulu scrambled out the door, each dragging three children. They piled into the family Peugeot, their belongings spilling out of overstuffed bags tossed in the trunk.

Ignatius looked on regretfully as his wife and children sped away along the main road toward the train station. Some of the homes along the way had been abandoned in the rush to escape, their front doors still open.

Within an hour, they were packed inside a train that would take them south through Kafanchan and Makurdi before crossing the River Benue into Igbo-majority Enugu. It was a full day's trek under the best of conditions, and these were not the best of conditions. The train was delayed at Kafanchan, as dozens of families dragged their suitcases, children, and livestock on board until there was no more space to stand. By the time they reached Makurdi, the overstuffed train refused to take on any more passengers.

The people of Jos understood suffering and death long before that mad scramble eastward as the pogroms exploded in 1966. The average life expectancy for a Nigerian born that year was just thirty-nine years. Ethnic violence flared frequently but many more died every day from disease.

My mother's family was lucky that death had not visited them up to that point, but they knew it was never far. Perhaps that's why, even in a train car filled with panicked Igbos flooding east, a part of Obiajulu still wondered whether they might have been overreacting. The sight of those bodies shook her, but maybe there was an explanation. Maybe the danger would subside in a day or two. Maybe they didn't have to abandon their lives, their schools, and their father. Amid the chaos, she held on to one silver lining: They were racing east, and east just happened to be the same direction as Arinze.

❧

My parents found each other one day after my mother arrived back in Enugu that fall of 1966. Obiajulu was clearing the yard when she looked out the front gate and noticed a handsome boy with a big smile leaning up against a pawpaw tree across the road. She screamed, loud enough for passersby to turn curiously, and ran to him. He held her tightly before speaking.

"I was so scared. I prayed for your safety, Obi."

"Things were getting really bad. We left as quickly as we could."

"You're safe now."

"We don't know where my father is. There's nowhere to hide in Jos. He's supposed to take the train here, but most Igbos on trains end up dead. If I lost him I don't know—"

"You're not losing anyone. Do you hear me?"

A week later, Obiajulu's father, Ignatius, did make it back. He'd fled Jos by train, just as his daughter had said, but she couldn't have known how close she came to losing him. Ignatius was traveling with three friends who'd decided to flee only after the northerners began hunting Igbos house-to-house, executing without a word those they found. The lucky ones were shot. Many more were beheaded. Ignatius hid in the bushes more than once to escape the violence and later emerged only to find his neighbors beaten to death with stones.

His friends wanted to take the fastest train home, but Ignatius knew the fastest train would be the most dangerous, an obvious target for bloodthirsty northerners lying in wait for Igbo passengers trying to escape. He preferred an indirect route that would take much longer but attract less suspicion.

They split up, just as many families and friends did in those dangerous weeks and months. Ignatius set off alone, getting rid of anything—his Western clothing, his bowler hat, his rosary beads—that gave away his Igbo heritage, and boarded a train heading west through Lagos. He'd lived in the north for several years and could speak Hausa fluently enough to avoid detection. Or so he thought. Halfway through his journey, a female passenger who'd been eyeing him suspiciously, stood up, pointed at him, and shouted in front of the entire carriage that he was an imposter, an undercover Igbo who should be searched and arrested. Ignatius froze in his seat, fearing the worst. Fortunately, there were no killers nearby to hear her.

His friends took the quicker route. He later learned they were shot dead in their seats less than an hour after the train left the station. It took Ignatius three full days to get back to his family. He arrived in Enugu hungry and shaken, but alive.

More than a million Igbos fled the north in those weeks and months, and by mid-fall, Enugu was overrun with an influx of new refugees, all competing to survive. Streets and markets were flooded with people, classrooms were overcrowded, and school dormitories were turned into makeshift shelters. Caroline arranged for her children to return to school as quickly as possible to create some semblance of normalcy, but there was nothing *normal* about Enugu that fall.

Arinze and Obiajulu crafted elaborate plans to see each other on weekends. And despite horrific violence to the north and the prospect of an all-out civil war growing, my parents' love story shifted into an important new phase. They spent so much time together, there was no need to write letters anymore. While chaos reigned all around them, their love was given room to grow. On weekends, they did their homework together at the well-stocked British library in Enugu, sitting on the floors between the shelves, whispering and giggling to themselves. On Sundays, he'd accompany her and her sisters to the stream, washing their laundry load in the water, while the group sat around and chatted.

Their secret love was becoming harder and harder to hide. More than once, Obiajulu's mother spotted Arinze on the street near their home and gave him a dirty look. He suspected she knew his visit to Rayfield the year before had been a ruse.

At one of the most dangerous moments in his country's history, the prospect of losing the respect of Obiajulu's family, which would threaten his future with her, felt like the most dangerous outcome imaginable. So, at barely seventeen, having already escaped genocide and with his country on the brink of war, Arinze decided it was time to stop hiding his love.

He sought out Ignatius at the market one Saturday afternoon and reintroduced himself.

"Sir, I'd like a moment of your time."

Ignatius raised his eyebrows coolly. "What is it?"

"Well, I'm sure you must know by now that I'm in love with your daughter. By the grace of God, I intend to marry her."

Ignatius took a step back to gather himself.

"You're sure about that?"

"I am. My only real worry is your wife. I'm afraid she hates me. I'm sure she would change her mind if she got to know me better. I'm respectfully seeking your help."

Obiajulu's father was taken aback by the young man's confidence. Arinze wasn't the most educated boy, but there was something about him that commanded respect. He was bold, and Ignatius secretly admired that.

"Come to our house for dinner tomorrow. You can make your case to my wife yourself."

He had two strikes against him before he walked in the door. Arinze was the son of a carpenter, poor even by village standards. No matter how charming or sophisticated or studious he may have been, he was in no position to give Obiajulu or her family the kind of success they aspired to. And beyond that, he'd lied. He'd invented a story to gain their trust in Rayfield and done God knows what with their daughter under their very own roof. A part of Ignatius appreciated the ingenuity Arinze had demonstrated by sneaking his way into their home—it showed real hustle—but Caroline was furious.

He arrived at their rented home on a hill the next day wearing his best shirt. Caroline met him at the door, and the look on her face made it clear his task would not be an easy one.

"Ma'am, thank you for seeing me."

She did not say a word. Instead, she kissed her teeth and sighed as she nodded toward the living room, where Ignatius was already seated reading a newspaper. The headline read something about Nigeria being on the brink of war if the Aburi Accords failed, but the rest of the front page was crumpled.

Arinze waited until Caroline was seated next to her husband, then launched into pleading his case.

"I was dishonest with you in Rayfield. And for that, I apologize. The truth is—I was desperate to spend time with your daughter. You must understand that I'm deeply in love with her. I would do anything for her—and for you."

His voice was strong and clear, carrying an authenticity that pierced Caroline's wall of motherly outrage—at least a little.

"Why should I trust you to care for my eldest daughter?"

She was subtly shifting the conversation away from Arinze's duplicity the year before, toward his limited career prospects.

Caroline and Ignatius had worked hard and saved up to educate Obiajulu in Jos, where the schools were better and the teachers came straight from England. In a country with no social safety nets, where people had little control over the turbulence of their daily lives—droughts, disease, ethnic violence—the one thing they could control was how much energy they invested in their children's future. Caroline wasn't about to let some village boy throw that all away. Arinze was poor; his dad was a carpenter who struggled to find work. *What kind of life could he offer their daughter?*

Arinze explained that he was excelling in the sciences and planned to study medicine. He would eventually become a doctor, using his ultimate wealth to help the village.

Ignatius folded up his newspaper and gestured for Arinze to sit down. Perhaps they'd written off the boy too quickly. Caroline's arms stayed crossed. Arinze turned to her.

"I know you may doubt me, but believe me when I say you can trust me with your daughter. I would never do anything to jeopardize her future, a future I hope includes me."

At that moment, an unsuspecting Obiajulu walked into the room, stunned to see Arinze talking to her parents in her own home for the second time in the span of a year. She almost fainted.

"Don't worry." Arinze stood up and approached Obiajulu. "I wanted to speak to them. They need to know my feelings for you."

Obiajulu's eyes widened in disbelief. *Maybe this boy really is mad*, she thought.

By the time Arinze left that night, they had come to an arrangement. He had permission to visit Obiajulu in their home, but nowhere else, and only when one of her parents was there. There would be no sneaking around, no unsupervised visits, no more lies. And they needed evidence that Arinze was doing as well in his studies as he claimed.

It was a grand victory for the young man. After more than a year of hiding and sneaking and lying, he had confronted and overcome the greatest obstacle in the way of a long and happy life with Obiajulu. With her parents' blessing, suddenly everything was possible. Had he been paying closer attention to the world around him, however, he would have—and maybe should have—predicted a far more dangerous threat than disapproving parents.

Within just a few months, the military governor of Nigeria's eastern region, Lieutenant-Colonel Emeka Ojukwu, formally declared independence. The Igbos in the east would split off from the rest of Nigeria and have their own country, the Republic of Biafra. Almost immediately, the place that had been a safe haven for 12 million Igbos, the place where my parents' love was born and growing, would become the epicenter of one of the deadliest civil wars in African history.

❧

Within months, the Nigerian government had ordered a blockade designed to starve Biafra of resources. No food was allowed to enter the new republic. Starvation was a cheap weapon of war and, as Biafrans would learn, one of the most effective.

Still, the Igbos' morale was strong at first. Our people were excited about the prospect of having their own country, especially after the horrors of the pogroms. But by early fall, federal troops were marching on Biafra, burning and killing everything in their path.

As rumors of an imminent invasion grew louder, Obiajulu became a refugee for the second time in her young life. She and her family, who frequently traveled between Enugu and Oghe, were now forbidden from leaving the village at all. She was safe at first, but the powerful passions of teenage love were soon replaced by an ugly fight for survival.

It was only a matter of time until Enugu fell. Once it did, Oghe's elders debated how much longer they should remain in the village as the civil war's bloody front lines crept toward them. Families consumed by fear began to split. Village elders who'd ordered people to stay watched in horror as their people fled.

Obiajulu's parents didn't waste any time. When the echoes of gunfire reached their compound one morning, they packed their children shoulder to shoulder with crates of live chickens and goats on stake bed trucks known as 911s, and trekked deeper into the heartland of the new Biafra, frightened, homeless, and hungry.

It was a gut-wrenching decision. The village was the heart of Nigerian culture. Leaving Oghe meant leaving a community that had supported their families for generations and venturing out into a new and dangerous world alone.

They sought refuge in towns farther east, where they rented rooms in other people's homes for weeks at a time before being forced to flee again. Ignatius was soon sent to help in the war effort and Caroline was alone again caring for seven children.

🖎

Obiajulu prayed for Arinze every day. She didn't know how close he was to the fighting, or whether he was still alive. She'd heard his family had traveled farther south to Umuahia, along with some relatives in the army, but no one seemed to know anything more. If Arinze was near the front lines with Biafran soldiers, he was in a precarious position. They lacked the sophisticated weaponry of the enemy, and most of their guns were leftovers from the Second World War. Obiajulu couldn't help but wonder whether, in the midst of his daily grind of survival, he longed for her the way she did him.

But Obiajulu had less and less time to indulge in teenage heartache as her responsibilities grew. Beyond caring for her siblings, she played a central role in making sure the family did not die of hunger as so many did.

Her job was to peddle garri in the local market of whatever town they happened to be in. Her mother would make the smooth yellow flour from cassava roots gathered by the children, and Obiajulu, the eldest, had to find buyers. It was the family's only real source of income in those days, and whatever money they earned went toward buying more food to keep them from joining the ranks of Biafra's starving.

Obiajulu was hawking the family crop one night at a market in Okigwe when she faced down death for the first time. Most of the markets in those days operated at night to avoid attracting unwanted attention from the aircraft that constantly lurked overhead. She was in the middle of laying out her goods when she suddenly heard the roar of engines. At first, she wasn't sure if it was one of *theirs*, but no one was taking any chances. Within seconds everyone in the market was scrambling to take cover; people ran and hid under wooden stalls and beneath palm fronds. Besides the sound of crickets, there was absolute silence.

And then it happened. Two bomb blasts shattered the peaceful evening commerce. A flash of fire flung Obiajulu to the ground. She was covered with bodies and body parts when she opened her eyes, her ears filled with faceless screams as the machine gun fire began to rain down.

She really thought she was dead. And being a thoughtful young woman, even in the afterlife apparently, she decided to lie perfectly still on the blood-soaked soil so her ghost wouldn't frighten the survivors. She lay like that, under the weight of death and innocence, until after the gunfire stopped. She would have lain there all night, perhaps, if her younger brother Eddie hadn't emerged and dragged her from the pile of bodies into the bush.

She took a deep breath, dusted herself off, and looked back at the carnage that was the marketplace. It was mostly women and children who lay bloodied and lifeless, people she had been speaking to just minutes earlier. The only noise in the grim silence was the groans of the injured pleading for help.

The family fled again that night, seeking temporary safety in another dying village. Despite the horrors of that day, and countless others, her mind always went back to Arinze. She wondered what he would make of her close brush with death, how he would comfort her and hold her close. That night, she reached under her sleeping mat and pulled out a crumpled photo. It was the only one of the two of them together. Arinze was sitting down, beaming brightly, looking up at her with reverence, as she smiled into the camera. She brought the photo closer, staring intently into his eyes, as though his face alone could somehow ease her suffering.

꙳

There were no more bombings right away, but the prospect of starvation was becoming real. There just wasn't enough cassava to sell. And even when Obiajulu had good days at the market, she never made enough to feed everyone.

At first, the family relied on donations from the International Red Cross, but they didn't come often enough, and at least one of their planes had been shot down. Relief trucks carrying food were sometimes hijacked and robbed. So, for a while, Caroline and her family of seven were reduced to hunting for termites, grasshoppers, and snakes to supplement the occasional cassava she received from generous strangers. No matter how little food they had on their plates, the children were taught never to eat *everything*. Not knowing when the next meal might come, saving half a grasshopper or a tiny slice of yam for the next day might be the difference between life and death.

The trick was not enough to save Arthur, one of Obiajulu's younger brothers, who at just eight years old joined the ranks of lost Biafrans. They'd managed to send word to Ignatius that his son was unwell, but the little boy died six hours before his father made it home. They could never be sure if it was the hunger or lack of medicine that got him—Arthur was asthmatic—but it devastated the family just the same when he couldn't wake up one morning.

The family packed up again later that day as rumors swirled that enemy soldiers were close. They filled the back of a 911 and headed for Olo. The chickens and goats were long since gone, and the children were down to six. As her mother wailed, the sound of artillery and gunfire reverberated in the distance. Obiajulu had a flash of desperation on the bumpy road, certain she would never again feel the safety and warmth that Arinze's presence always brought.

They stared hopelessly as they passed through dying towns where stray dogs dug through trash heaps, doors clanged open in the breeze, and children lined up for meals at makeshift Red Cross stations, meals that might have been their first in days, weeks—or even their last. She overheard the driver say that all the talk of Biafrans winning the war was rubbish, that the war could drag on for several years if Biafrans didn't do the smart thing and surrender. She began to fear she would never see Arinze again.

It was a dark night under an ocean of stars for a family on the brink. The truck broke down at least twice, and they were terrified of the constant buzz of military aircraft hunting for Igbos. The driver stopped in the small hours of the night at a blown-out bridge three miles from their destination, forcing them to scale a jagged ravine by foot in the pitch-black, dragging toddlers and their dwindling possessions to the other side. The sun was rising as they came across a barn on the village outskirts, where a friendly family let them rest.

Within a few days, they'd somehow scraped together enough money from hawking their wares to open a makeshift restaurant in the barn where they slept. Obiajulu helped cook as her younger siblings waited on customers and cleaned. The house special was a variation of pounded yam and ogbono soup, and the scent of homecooked food during a famine drew a steady stream of grateful Biafrans. The money they earned meant there would be no termites or crickets for dinner, at least for a while.

Obiajulu was focused on chopping okra one day when a smiling young man settled at a table in the back of the barn. Eddie, too, began to smile when he turned to take his order. He called out to his big sister to say hello to their newest customer.

She screamed.

He'd found her. In the midst of a civil war surrounded by death, Arinze had found her.

She sprinted toward him and jumped into his arms, somehow not spilling the order of okra she was carrying.

She could only say one word at first: "How? How? Just how?"

They snuck off that evening and hid away in an abandoned church. Leaning up against each other in one of the pews, he explained he'd spent three weeks trying to track her down. He nearly had them in Otuocha before the trail ran cold, then in Okigwe before a family friend said he'd just missed them. With every reason to fear the worst, he somehow never doubted they'd be reunited.

She told him about her own brushes with death, about Arthur, about thinking she was a ghost, about eating insects, and about late-night hikes through the bush.

He wrapped one arm around her shoulders, and, with the other, dipped into his pocket and pulled out a faded gold ring.

"It was my grandmother's. Now it's yours. Be my wife."

She said yes with a gentle kiss and began to cry.

"I can only stay another day," he told her. "Just know that when this ends, and even if it doesn't, I will find you and we will start a proper life together."

Less than a year later, the Biafrans surrendered and the fledgling nation was no more. Roughly 100,000 soldiers were killed on both sides, but the civilian death toll was far worse: an estimated 2 million died from starvation in what was the continent of Africa's deadliest recorded famine at that time.

Arinze was true to his word. He eventually tracked down Obiajulu back in their shattered village, Oghe, a few weeks after the surrender. And within another two years, they'd set off to Europe together.

My parents had survived ethnic cleansing, civil war, and the continent's worst famine, living the kind of trauma that neither my children nor I will ever truly understand, and yet their hearts stayed open—to each other, at least. My mother taught me so much about success in this world, so much about work and discipline and sacrifice, but her ability to stay in love with my father through unimaginable suffering may have been the greatest lesson of all.

Chapter 8

There are no artifacts from the Biafra war in the Imperial War Museum in South London, a green-domed building that sits behind a pair of 15-inch naval guns. The large, airy spaces are devoted to proof of the military strength that supported colonization, not the ugly fallout often left behind.

Still, the museum was one of the few British cultural sites my mother dragged me to on the rare occasion she had a free weekend afternoon. I usually resisted, but secretly, I loved those moments together, learning about the world and then demonstrating my newfound knowledge with the quizzes that almost always followed.

"Who was the first woman to be given a George Cross?"

"Odette Sansom! She was a secret agent."

"What was the purpose of the Treaty of Versailles?"

"To end the First World War!"

I was only nine years old, but she had trained me well. I blurted out the answers with smug satisfaction, eager for the next test. But somewhere between my beaming smile and my mother's unabashed pride, I remembered what I had been trying to forget.

It was the end of summer, and in a few weeks, I'd be going back to school. While I'd become a good student, school also meant water. Water meant shame.

Every Wednesday, my school held swimming lessons at Crystal Palace Sports Centre, a giant recreation complex in South London. And almost every time I entered the pool, the other girls would scream and cry, fearing the blackness of my skin would turn the water brown. They'd shout out, "She's coming!" as I approached, thrashing and splashing toward the safety of the shallow end, if they didn't jump out of the pool altogether.

I did everything a little girl could to avoid those swim classes. I cried, I faked sickness, I hid in bathroom stalls. I would have sold my soul to avoid ever seeing those girls again, but my mother was unyielding.

She hoped our little cultural excursions and day trips would help shift my focus. We'd wander inside those art galleries and museums, and she'd test me on an obscure date or event. We'd argue over minutiae. Then I'd hear the sound of children playing or girls laughing, and the dread would fill my chest.

"What's the matter?"

"Nothing."

"I hope you're not thinking about those girls."

But those girls were all I thought about. She'd try and turn my attention back to an exhibition or an artifact, but there was nothing she could say to make it better.

"Are you going to let them ruin the whole day? They're not even here."

My mother tried to empathize, she really did, but she never could understand the depth of my struggles. We had been raised in different worlds. Her school years were marred by hunger and homelessness. She'd faced down the prospect of death almost on a daily basis. Dealing with a few mean girls would have been a welcome challenge.

Besides, my mother grew up in a place where people did not talk about their problems—they took action. They fetched and hunted and ran and fought. But in London, my problems were often related to things like social acceptance, hurt feelings, and loneliness—things that didn't carry such simple solutions.

Still, she desperately wanted to make it better. After I'd gone to sleep that night, she dialed all thirteen digits of my grandmother's number on the rotary phone in her bedroom. She needed some advice. Was she being too hard on me? Too tough? What else could she do? Clearly, her efforts weren't enough.

The conversation with my grandmother was short. After less than ten minutes, they'd devised a plan.

I was in my room studying one night when I heard the jingle of her keys on the front step. The front door creaked open, there was a rustling of papers as my mother checked the mail, and the clank of her heels as she climbed the stairs. My bedroom door swung open. She dropped her purse onto my bed and plopped down beside it, looking at me with serious eyes. It was cold. It was always cold in our house. I could see her breath in the air when she spoke.

"I want to talk to you."

I put down my fountain-tip pen and sat up in my desk chair. She took a deep breath.

"I'm sending you home."

It took me a few moments to process what she meant. *Home* in my family could only mean one place: the village in Enugu where my mother grew up. It was a place with no running water or electricity, where cockerels, goats, and stray dogs shared the same bumpy roads as rusty Peugeot 504s, a place where people sometimes caught their own food and boiled their drinking water over kerosene stoves on the dirt floor. That was home.

"Why?"

"Training."

Her tone seemed more regretful than decisive. In the West, the word *training* implies an event one is being prepared for—a marathon or a tournament perhaps. Among Nigerians, *training* is different. It's a period during which children are taught to practice dutiful obedience, to stretch themselves through unrelenting discipline, to survive on their own. As part of our *training*, Nigerian children born and raised in the West are sent to live with extended relatives in Nigeria before the end of high school. The thinking goes like this: If a child can withstand the harsh conditions of the village, such as walking miles to collect their own drinking water, washing their clothes by hand, or living without electricity, they can pretty much withstand anything in Europe or the United States.

My mother decided that words alone would not help me overcome my adolescent challenges. A few years in Nigeria, living with *her* parents, would teach me what she could not.

This practice, known among Nigerians as *shipping back*, has been part of our culture for decades. Almost every Nigerian knows someone who was shipped back. One of my cousins was shipped back to Lagos when she was ten. Mrs. Maduka's daughter, who was always hanging around older boys, was shipped back to Anambra when she was fourteen. A girl in my school with round-rimmed glasses and crooked yellow teeth had been shipped back the month before.

I don't remember being surprised when I heard, but somehow I never imagined it would happen to me. I could feel the teardrops pooling as she dictated my fate.

"When?"

"In about a month."

The quick turnaround made it feel more punitive, as though she'd forgotten *I* was the victim. I had only a few weeks to say goodbye to everything I knew: my siblings, friends, teachers, the VHS tapes in the cellar with my favorite episodes of *Tom's Midnight Garden*.

"How long for?" I asked.

Training usually lasts two to three years, but three years can easily morph into a child's entire high school career. My American cousins were shipped back to Nigeria when they were children and didn't return to Texas until they were in their late teens.

"However long it takes."

She was resolute, having decided that it was in my best interests to live under intense discipline in one of the most remote parts of Nigeria for as long as necessary.

In Britain, it was much harder to discipline children. There were too many distractions. Even if parents tried to replicate the tough-love Nigerian principles of hard work and respect, a culture clash often emerged, pitting the strict Nigerian values *inside* the home against the more relaxed Western values *outside*.

Nigeria, for all its faults, was the perfect place to toughen me up. It was an elite training ground for resilience; the West Point academy of perseverance. Survive in Nigeria for ten years and you can survive anything. *Thrive* in Nigeria and you can change the world.

I woke up with a throbbing headache the day I was supposed to leave. The tears started even before I said goodbye to my siblings, who had lined up by the door. Everything about that morning felt wrong. Most Nigerians were trying to come *here*; why on earth would I be going *there*?

I had traveled to Nigeria only twice in my life, but I knew all about its reputation. Even back then, Lagos was not for the fainthearted. I'd overheard my uncles talk about the violent crime and rampant theft, how drivers locked their car doors whenever they were stuck in traffic to stop armed robbers from climbing in. And the corruption and the lawlessness were some of their many punch lines. I couldn't believe I was being *shipped back* to such a place. But I knew there was no getting out of this.

The car ride to Heathrow along the M4 was both frenetic and calming. I gazed out the window at the world whizzing past, eager to collect every memory I could of the world I was leaving and hopefully would one day return to. I stared at the River Thames lapping against its banks, the lollipop men with fluorescent yellow jackets helping schoolchildren along zebra crossings, the red-roofed terraced homes that littered Hounslow. I tried not to cry.

We arrived at the airport early. My mother's eyelids sagged. She'd been up all night wrestling with her decision. With her head down as though in mourning, she reached for my hand and led me through the double glass doors, through the crowded departure hall, and up to the British Airways counter. My mother handed over my passport and dumped my suitcase on the scale.

I barely had time to hug my mother goodbye before one of the agents whisked me off toward the security area. As the distance between us widened, I had an intense urge to break free and run back to her. I wondered if she felt the same way. I kept looking forward, though. I was nine years old. I had to prove I was a big girl now.

↜

Murtala Muhammed International Airport in Lagos was its own brand of beast, teeming with colorful characters and the hot hum of movement. Everyone seemed to be yelling and waving their arms. And everything was broken. Conveyor belts at baggage claim lay still, letters from neon signs hung unlit, fans sat idle. I wandered through the hurricane of activity bewildered and alone, except for the white flight attendant leading me toward the exit. This was Lagos, raw.

It took me a few minutes to realize my own foreignness. I was technically among my people, but really, I was the onlooker now, the confused spectator. My clothing, my walk, my smile, my curious gaze—everything betrayed my otherness. The realization made me squeeze the flight attendant's hand even more tightly. In that moment, the two of us had everything in common.

Just as we were about to step into a more chaotic scene outside, Grandpa waved us down from the other side of the arrivals hall. He was slender-framed, with a well-fed stomach.

He approached me, arms wide, and squeezed tightly.

"Good afternoon, sir," I said as I stepped back.

As I had learned years earlier, in Nigeria, *sir* and *madam* are required terms of endearment for our elders, especially those in our own family. We greet our loved ones in the same way others might greet a teacher or police officer. It's one of the small marks of discipline that define Nigerian parenting. If we show respect to our elders with the small things, we're much more likely to do it when the stakes are higher.

Grandpa nodded a thank-you to the flight attendant and bent down to take my hand. Sensing my unease, he chimed: "It won't be as bad as you think. The air is fresher in Enugu."

We didn't talk much during the car ride to the local airport. He wasn't the sort of person who navigated light conversation easily, and I was distracted by everything I saw. Cars swerved onto the sidewalk to cut around traffic and people seemed to be collecting rainwater in buckets on the side of the road.

In a flash, we were boarding a small plane en route to Enugu. Even as a little girl, I could tell that our plane was not the best in the fleet. We sat down in the emergency exit row, and the flight attendant explained the emergency procedures in frighteningly vivid detail.

"If the plane crashes and catches on fire, open this door to jump out. But wait until the plane has completely stopped so you don't injure yourself."

It sounded like she was speaking from experience.

We arrived in Enugu just before dusk. Grandpa was right. The air was easier to breathe. My uncle was waiting in a red Mazda, and soon we were speeding into more familiar landscape. Enugu is a hilly city, known for its coal-mining roots, where modest concrete buildings dotted the lush landscape connected by dirt roads.

I knew that Enugu had a lot of fancy areas, but my grandparents' neighborhood wasn't one of them. We pulled up next to a two-story house that was a block away from a landfill. A stray dog lingered out front. Through the open door, I glimpsed bare walls, broken tile floors, and a tattered sofa. The place was in such disrepair, I wondered if a would-be thief might actually be tempted to leave something behind for us.

My grandmother, the same powerhouse matriarch of the family who cared for us in London after my father's death, emerged and greeted me with a rare smile. She was slighter than I remembered, but just as fierce when she needed to be. She yelled out to one of my cousins to take my suitcase upstairs, and before I could even ask for water, she sent me straight to the kitchen to help prepare dinner. My mother had given my grandmother strict orders to expose me to the unfiltered demands of Nigerian life as quickly as possible, and she wasted no time.

Maybe some nine-year-olds know their way around the kitchen. I didn't. Back in London, my mother insisted she would handle all the cooking so I could devote my time to my studies. She joked that she'd give me a crash course on meal preparation a few months before my wedding in, say, twenty years, but until then, I would spend my evenings reading. That left me barely knowing how to use the microwave. And I had certainly never worked in a kitchen like the one in my grandparents' house.

The cream walls were soiled with soot and sprayed food particles. There was a small kerosene stove on the floor. Apparently, they cooked while sitting on the ground.

Chinyere, the house girl, appeared and introduced herself. She was burly and thickset. I asked what we were having for dinner.

"Is that how you greet me? You don't remember me?"

I shook my head, no.

"I'm your grandma's brother's wife's cousin."

In Nigeria, everyone is family.

"You eat pigeon?" she asked.

She led me outside to the backyard, placed a washbasin facedown in the middle of the small clearing, and propped it up with a stick. She then sprinkled some grains of rice under the basin, attached a rope, and strung it back to the house, where we waited by the kitchen window in silence. Eventually, two pigeons swooped down and started pecking the rice. Chinyere yanked the rope, and the basin fell with a clang. This was dinner. She saw the shock on my face and burst out laughing. Born and raised in the village, she'd mastered this trick as a child. Within forty-five minutes, she had plucked those pigeons, fried them up in mmanu, and served them with boiled plantains.

The lights went dark just as we sat down to eat our pigeon. My grandmother laughed as she lit a candle and explained that electricity in Nigeria was intermittent. Like a rebellious teenager, it sometimes stayed for an hour or two before disappearing for days without notice.

NEPA was an acronym for National Electric Power Authority, but everyone joked it stood for Never Expect Power Always. Once every few days there'd be a synchronized groan of frustration as the whirling fans would stop and everything would go dark. I learned to avoid games and activities that relied on electricity. TV was out of the question. With no other reliable options, I soon began to covet books and my grandparents' stories.

On my first Saturday, I woke to the sound of bleating goats, but the neighborhood was otherwise silent. We lived just off a busy road in Abakpa Nike where there was usually an orchestra of horns and engines fighting for attention. Not that day. I got dressed and went downstairs.

"Mama Nnukwu, why is everything so quiet?"

"Last Saturday of the month. Go an' get a broom."

Back then in Nigeria, the last Saturday of the month was Environmental Sanitation Day. Every Nigerian was required to spend the morning cleaning, scrubbing, and polishing their home. It was so rigorously enforced that anyone caught walking or driving on the streets when they were supposed to be at home cleaning could be fined or worse. The architect of this policy was the former head of state Muhammadu Buhari, a military commander who believed that cleanliness was the simplest way to instill discipline in society. Just the small act of cleaning out gutters or cutting grass would teach children the importance of routine, healthy habits, and purpose.

The law stipulated that only mornings would be dedicated to cleaning, but in our house, the tasks consumed the entire afternoon. That first Sanitation Day, Chinyere and I scrubbed the floors, swept outside, and dusted cobwebs from the ceilings. By late morning, we were filling a massive basin with hot water, blue powdered soap, and dirty clothes to do laundry. By early afternoon, we were cleaning toilets.

This was no simple task. In most Enugu homes at that time, the toilets didn't flush, so we were forced to pour buckets of cold water down the bowl and scrub for as long as we could hold our breath.

Generations earlier, the British colonial system brought life-changing amenities such as toilets, drainage systems, roads, and railways, but after Nigeria won its independence in 1960, we never learned to maintain them.

Without working faucets, we relied on a dirty, muddy stream known as Mmili Ocha for all our water needs. On weekends, Chinyere and I would carry large green buckets a mile to the banks, fill them with brown, murky water, and balance the buckets on our heads for the bumpy journey home.

I was learning that childhood in Nigeria was truly a crash course in survival. There was no time for play or exploration. As soon as children could walk, understand simple instructions, and carry out basic tasks, they were expected to contribute to their family and community. Nigerian children fetched water when they were barely bigger than the buckets they carried on their heads and so would I.

So much of what I learned in those initial days was about endurance. I washed and cooked and scrubbed the floors. But nothing was more important to my grandparents' sustenance than making sure I returned home with my buckets full. That was our family's drinking water. The pressure to defend my precious cargo was real and intense.

Chinyere initiated me in the art of bucket carrying.

"Never ever turn your head! Keep it straight the entire way. Glance down every few seconds to make sure your path is clear."

That first time, I balanced the bucket with frightening precision, almost afraid to breathe. My teacher carried one large bucket on her head and a smaller one in each hand. I had the luxury of being able to use my free hands to steady the bucket at first. Still, I barely made it. The sun pounded down on my back and the sweat stung my eyes. The walk lasted an eternity. For a newcomer like me, a trip to the stream and back could swallow an entire afternoon.

Every so often, the state government would bless us with running water for a few hours. Those days were almost better than Christmas. I remember the neighbors cheering in the streets and Chinyere deliriously screaming, "*Pump n'agba*," or "Water is flowing," as she ran across the yard, arms spread wide with gratitude. You'd think Nigeria had just won the World Cup. We'd crouch down next to the faucet, collecting gallons of running water in jerry cans until they were all full.

But those *running water* days were rare, and it was often on Saturdays, after a long day of cleaning and fetching water, that I missed my old life the most. I'd crawl into bed soon after sundown, squeeze between my grandma and Aunty Nneka, and stare at the ceiling, praying with all my heart that I'd wake up back in London, back in my cold bedroom, back at my elementary school, answering questions about our weekly reading assignment in Mrs. Patterson's English class. I wondered what my brothers were doing, how the dinnertime book club was going, and whether the kids at the pool were having more fun now that I was no longer a threat.

Just after Christmas, my aunty Ngozi showed up at Grandma's house for a visit on her way back home to London. Her accent, clothes, and perfume reminded me of everything I'd left behind. She asked if I had anything to give to my mom in England.

"No."

"Why don't you write a letter?"

"Now?"

I grabbed a sheet of paper. With only a few minutes to scribble out a message, I let my desperation take the lead. This was my rescue appeal. I'd better make it good.

I hate it here. Please, please take me back. Why did you leave me? Why is everything always broken? I want to come home. Please!

༕

After a few weeks, it was time to go to school.

There were no swimming lessons here, but I knew I'd still be an outsider. My mother had enrolled me in a small elementary school next to a university campus. I wasn't sure what to expect, but it was clear the rules here would be vastly different from anything I'd experienced in London.

My uniform was a red and white checkered dress with soft fabric and dark red buttons. I secretly tried it on three times the day before school started, excited for any excuse to take a break from cleaning.

My school was made up of a series of aging one-story bungalows on the edge of a barren field. The classrooms were bare. Without children in them, it would have been hard to guess they were part of a school. Mrs. Okonkwo was the main teacher for my year. She swayed her hips as she walked up and down the aisles, waiting to catch the kids misbehaving. Her singsong Nigerian accent was thick but commanded respect. A cane lay ominously across her desk; a prop, I thought at first, to scare the students into submission.

I learned on my first day not to underestimate her. As I would later find out, parents in Nigeria were so keen to churn out well-behaved kids that schools offering corporal punishment were considered helpful allies. The cane was not a prop.

I was on my best behavior the first few weeks, but one morning when she stepped out, I started chatting with my new classmates. Without warning, she burst into the room, demanding to know who'd been speaking out of turn.

She looked directly at me: "Oyibo, was it you who spoke?"

Oyibo had become my nickname. It meant "foreigner" and was generally not a compliment. Most Nigerians viewed foreigners as naïve. They were people who lacked survival instincts, weak people who couldn't do things for themselves. Every time someone called me Oyibo, I remembered why I was in Nigeria in the first place.

"Eh, Oyibo, was that you?"

"No, not me."

"Oya Oya, if you don't speak de truth, you'll see what I will do to you." She grabbed her cane from her desk.

"Fine, it was me. Sorry, madam."

The admission was a mistake. She beckoned me to the front of the room. When I reached the blackboard, she told me to face the class and raise my arms in the air. I followed her orders sheepishly, thinking it was some kind of joke. She warned me not to drop them until she said so. That was my punishment—holding my arms high above my head, even when they began to ache and even as the blood rushed down toward my shoulders. I winced in discomfort after a few minutes and glanced at Mrs. Okonkwo, who was casually flipping through a beauty magazine. I tried to lower them, hoping she'd have mercy on me.

"Did I tell you to drop them? If you drop your arms before I say so, you will clean the toilets-oh."

Cleaning toilets in a school with no running water was one step above corporal punishment. The pipes and bowls were constantly blocked with brown sludge and slime, and the floors were drowning in urine. The smell was nauseating. The threat of bathroom duty gave me the jolt of energy I needed to hold my arms higher. From that moment on, I became the quietest child in class.

My best friend in the early days was Okechukwu, a nine-year-old boy with a wide gap between his two front teeth, who lived across the street from my grandparents'. He explained how to navigate Mrs. Okonkwo during one of our car pools home together.

"You can't relax with dat one. The minute you get comfortable she will turn on you-oh. Dis no be like your schools in London."

He offered to carry my backpack from his house to my front door one day, and I let him. Unbeknownst to me, my grandmother was watching in silent judgment from the window upstairs. He'd barely said goodbye when my grandma appeared behind the front door, shaking her head in disappointment. She yanked me into the hallway.

"It's time to take you to Chinedu."

Chinedu was a stylist at the hair salon off Apakpa Nike Road. Back in London, my mother always wanted me to have thick, healthy hair. But in Enugu, girls were expected to shave off all their hair before middle school so they could focus more on schoolwork and less on appearance or boys. The incident with Okechukwu prompted my grandmother to intervene early.

"Boys are no good," she warned.

I remember the empty feeling of seeing my hair, clumps of tightly coiled black cotton wool, falling to the floor as Chinedu severed my soft puffs from their roots. I looked up at the mirror when he was done. My reflection made no sense. If I had been born a boy, I'd probably look better than this. Nigeria had taken everything from me: my friends, my family, my free time—and now, my hair. I wondered what the girls at school in London would say now. *How loud would they scream?*

That evening I asked my grandma if I could call home. She rolled her eyes and exhaled loudly. Calls abroad required serious effort. Our landline could barely handle local calls, so calling from home was out of the question. She drove me to a local business where you could pay to call abroad. There was zero privacy in the small space, and while I didn't love the idea of a group of strangers listening in, I didn't have a choice. I gave the owner a card with my mother's number. He began dialing. My grandmother pointed at the clock, signaling she would only pay for five minutes. No more.

"Hi, Mum. It's me."

"What's wrong? It's late."

"I want to come home."

"I know. I got your letter."

"They made me cut my hair. I look like a boy."

"We can talk about this later. Where's Grandma?"

"I don't fit in. I don't have friends. I hate it here."

"Get some rest. I'll call you tomorrow."

My grandma motioned for the phone. She exchanged a few words with my mother in Igbo dismissing my concerns and hung up. We drove back to the house in silence.

I was beginning to understand that there was a different set of rules in Enugu, and no one was going to help me break them. If my grandma decided to cut my hair after catching me talking to a boy, no matter how innocently, her decision was cemented into law. There was no appeals process, no teacher or friend or parent to intervene. Children were expected to know their place, which was a place of unquestioned obedience.

This is how it was in Enugu. The rules applied even to the smallest things. I remember going upstairs one afternoon to read in our shared bedroom after an exhausting trip to the river. Just as I began to doze off, I heard my grandma calling my name faintly as if it were a song. When I ran downstairs, she asked me to hand her the glass of water sitting on the side table next to her. I looked at her quizzically. The water was obviously within her reach, but for my grandmother, it was perfectly reasonable—expected, almost—to wake me up from a nap, call me downstairs to hand her something that was already right next to her.

This type of discipline was all my grandmother knew. She was born in our village in the 1930s, at a time when childhood was all about re-silience, strength, and endurance. There were no toys scattered across her bedroom, no playdates with other children her age, no birthday par-ties. In fact, birthdays were barely acknowledged at all and most peo-ple used wars, droughts, or the death of a public figure, among other events, to remember the year their children were born. After school, my grandmother fetched water, cooked dinner on the dim glow of logs outside, bathed her younger siblings, and prepared them for bed. For all intents and purposes, Mama Nnukwu had been practicing the art of backbreaking endurance since she was seven, and that was exactly what I was there to master.

As the months passed, it became clear that no one was coming to rescue me. If I was ever going back to London, it would probably only happen after I'd learned the lessons I was there to learn.

Soon, a new routine began to emerge. I'd wake up, make my grandma pap and akara. Then I'd fill a bucket with hot water and fresh soap for her bath, and set aside a chewing stick so she could *brush* her teeth. On Sat-urdays, I was in charge of clearing the yard. I'd take my aziza and sweep away the grains of rice, fowl feathers, and specks of Omo soap sprinkled across the compound. I even stopped resisting toilet cleaning. I'd spend a good twenty minutes once a week on my hands and knees scrubbing the porcelain bowl back to the color it was supposed to be. Enugu had given me a taste of what my mother and grandmother had grown up with their entire lives.

My grandmother beckoned me from upstairs one afternoon, and I assumed she needed a tissue or a drink. Instead she told me to pick up the phone. It was my mother. It'd been so long since I'd heard her voice, I almost didn't recognize it at first.

"How is everything there?"

"Better. Thank you."

"Is there anything you need? I can send you those Breakaway chocolates you like."

For the first time, I didn't feel compelled to beg her to let me come home. I didn't even feel like crying. London felt like such a distant memory, I almost couldn't remember what it was like to have dependable electricity or clean, running water. Enugu was what I knew now.

"You haven't called home in a while. Is everything okay?"

"Yes."

"Good. I just wanted to check in."

"Love you."

"Me too. Bye."

My grandmother was staring at me when I hung up the phone. She scanned my eyes, expecting some sign of sadness after speaking to my mother. She seemed confused when I calmly said good night and walked upstairs to bed.

By that time, I had managed to win over Mrs. Okonkwo. I had a desk right next to the blackboard, and whenever our eyes met, she'd bless me with a smile. I refused to talk out of turn in class, even when tempted. I handed in my homework promptly and used every chance I got to show off my Igbo. Phrases like *Ke kwanu?* (How are you doing?) and *Daalu* (Thank you) became part of my everyday speech.

My London accent evolved into something much more Nigerian, and the Oyibo references started to feel more friendly. The other kids wanted to play with me during lunch break, and Mrs. Okonkwo relied on me for little things like distributing class materials and collecting homework. Eventually, she made me class prefect, which meant I was in charge whenever she left the room.

Unlike my school in London, where I couldn't even get into a swimming pool without being shamed and bullied, I began to feel respected.

This was my second chance, a chance to have the kind of community and acceptance I'd always craved.

But it wasn't a perfect transition. The power of prefect status quickly went to my head. I'd strut down the halls shushing noisy classmates and writing down their names to share with the teacher. I saw myself as Mrs. Okonkwo's enforcer. Not everyone appreciated it. A few months into my reign, one of my classmates approached me from behind during playtime.

"Oyibo."

The voice was calm but angry.

"Oyibo, you no hear?"

I turned around. It was Kofi, a stocky Ghanaian girl I'd reported for talking too loudly earlier that day. Who knows which one of Mrs. Okonkwo's punishments she had endured?

"You told Mrs. Okonkwo I was talking?"

"Well, weren't you?"

She stepped into my space, her face only inches away from mine when she answered: "Watch yourself, Oyibo."

My heart fluttered. It was a direct threat, and I could tell she had the willingness and desire to back it up. But I was too busy to live in constant fear, and the next Saturday afternoon, Chinyere and I made one of our regular runs to Mmili Ocha to fetch water. The banks were crawling with people when we arrived. I'd been making water trips for more than a year, and I'd never seen it so crowded. Dozens of teenagers and children were sitting along the muddy basin collecting water, some playing, others washing their hair or clothes. It looked like an outdoor party. It was the heart of the dry season and there hadn't been any rain in months. The people were relying on the stream more and more for comfort and sustenance.

I knelt down along the crowded banks and filled three green buckets with water. In my early days, I was lucky if I could carry just one, but I had become a water-carrying pro. It was all about balance. I would place the largest bucket on my head and balance a smaller jerry can in each hand. I couldn't walk as fast as Chinyere, but I could go miles that way.

But this time, as I cautiously lifted my sloshing cargo, a familiar voice caught my ear. I slowly turned to my left, and there, just a stone's throw away, was Kofi. Our eyes met. Neither of us spoke. I shuffled up the bank as quickly as I could and called for Chinyere to follow.

We soon made it to the dusty road, where Chinyere overtook me. Usually, I liked it when she walked ahead. She artfully navigated every bump and pothole in a way I never could. But on this day, I didn't want to fall too far behind. I scurried to keep up, but within a few minutes, she was far off in the distance. That's when I heard it.

"Oyibo!"

Kofi had crept up behind me. I desperately looked ahead and realized Chinyere wasn't even within shouting distance. I winced. Chinyere had taught me a lot about carrying water, but she never taught me how to balance three buckets and fight at the same time.

I turned around, and Kofi, having dropped her buckets, lunged at me. I jerked backward, and a splash of water hit my face, as the big bucket teetered.

"Leave me alone. I don't want to fight."

I turned and walked away, all three buckets still safe. Thinking a miracle had occurred, I smiled. *Had I really just stood up to her?*

That's when she pushed me from behind. The bucket on my head flew off and landed with a crack. It bounced and rolled away, leaving an entire week's worth of water forming a puddle at my feet.

Kofi erupted in laughter. I dropped the jerry cans and attacked. She was much bigger. I clawed at her arms and face. She grabbed my neck.

We wrestled and ripped, neither willing to let go, until we tumbled into the mud. She pressed my face into the wet earth, but I rolled away, and we both jumped up. I took a step back and kicked her in the shin as hard as I could. Her leg buckled. I did it for the water I spilled, the water my family would go without that night, the water back at the sports center in London where my Blackness scared the other girls.

Kofi let out a yelp, scrambled to her feet, and scurried off. I gathered my empty buckets and raced after Chinyere. I didn't tell her about the fight when I finally caught up.

"Chinyere, they broke."

"What did?"

"The buckets."

"What?! Didn't I teach you how to balance?"

"I'm sorry."

Chinyere rolled her eyes and sighed.

"There's not enough water. I'm sending you back tomorrow."

I don't know if that was the moment my mother had wanted for me when she shipped me back. I had been forced to fight—to fight for my family and to fight for myself. I don't believe there's a real winner in any fight, but sometimes, you fight anyway. You drop your buckets and close your eyes and kick and claw and scream with everything you have. Even when you're terrified, or alone, or the only Black girl in the pool, you fight. To be Nigerian, I was learning, was to fight.

My mother had spent the first decade of my life trying to teach me that lesson, a lesson as old as my country. But it took a broken bucket on the banks of the Mmili Ocha for it to sink in.

My mind was still buzzing from the fight when my mother called that night. I talked about my duties as a prefect, the near perfect grade on my recent math test, and the latest batch of egusi soup I'd cooked, but I didn't tell her what happened at the river. I knew she couldn't help me even if she wanted to. She was a continent away with a mountain of responsibilities of her own. These were problems I'd have to solve alone. As the conversation continued, she sensed that I was distracted.

"I don't hear from you anymore. You've been busy helping Grandma?"

"Yes. There's a lot to do here."

"You doing okay?"

"Fine."

"I miss you, my daughter. I've been thinking. Maybe it's time for you to come home."

She paused, expecting me to scream with joy. I did not.

"You can continue your schooling here," she said. "It's been long enough."

"But—I'm not sure . . ."

"Start gathering your things."

A return ticket to London was supposed to be my reward for mastering life in Enugu, for chasing pigeons, for serving my elders, for winning the fight. But after almost two years, London's appeal had faded. I'd learned how to hold my own on Sanitation Days, I appreciated the simplicity of my shaved head, and I could walk for miles without spilling a drop of water from my buckets. I had been *trained*, just as she wanted me to be. *What use would I be back in London?*

"How long do I have left?"

"Until the dry season ends."

Chapter 9

My mother wasn't interested in movies when she first got to England. And even if she was, *Live and Let Die* wasn't one she'd ever have chosen as her first Hollywood feature. She knew the movie was provocative, that there were car chases, sex scenes, and violence. Everything she'd heard warned of drug dealing, witchcraft, and shoot-outs, and those certainly weren't images her mother would approve of.

In Nigeria, the only films she'd ever watched were the colonial propaganda shorts projected onto sheets in the village square; movies that featured a man who didn't talk, named Charlie Chaplin; or the ones that offered instructions on how to navigate life in London. She knew *Live and Let Die* would be a whole different experience.

But Obiajulu was on a mission. At twenty-one years old, she was the oldest student in her biology and math classes at Walthamstow Forest, the East London high school she'd just enrolled in, and the only pregnant one. After almost a year in a new country, Obiajulu was running out of strategies to make friends. She would have done anything—and everything—to forge a connection with the giggly, carefree sixteen-year-olds who filled her classroom, and she needed some shared experience, some piece of pop culture, a concert or even a song, to prove she was just like them.

New experiences are hard enough for anyone. But the transition from Enugu's smoldering ruins to Walthamstow's warm, furnished classrooms was more than a new experience—it was like changing planets.

That first day, Obiajulu walked into the classroom of chattering teenagers, nervous to be among the kind of real-life British people she'd read about in books. She was far too self-conscious to introduce herself to any of them directly, too afraid that her village ways might make her look odd or naïve. Instead, she kept her head down as she shuffled to an empty desk in the back of the room. She waited for someone to notice her. No one did.

On her second day, Obiajulu decided to take a risk. She asked the girl sitting in front of her to lend her a sheet of paper. She prayed the simple gesture might spark a conversation.

Of course! Take as much as you need. What's your name, by the way? she hoped the blond girl would ask.

Thank you so much! Call me Obi. I've only been in England a few months, but it's lovely so far!

Oh, that's brilliant! You should join us after class. A few of us are going to the pub. We can show you around.

That's not what happened, of course. The girl smiled, handed Obiajulu a sheet of paper, and quickly turned back around. It was over in a flash.

After class, she packed up her belongings and moved toward the door as slowly as she could, hoping to fall into one of her classmates' conversations. It didn't work. No one seemed to see her. And even if they had, she had a hard time understanding the bits and pieces of conversation she picked up.

Who is this queen they're talking about—one that makes music? The only queen she'd learned about in school was Her Majesty Queen Elizabeth.

This was all so different from Enugu.

Back home, Obiajulu didn't have to trick people into talking to her. She was one of the popular girls, the one the other students circled around, along with her best friends, Petronila and Nneka. In England, she was invisible.

At first, she tried to focus on the bright side. Attending school, any school in England, was a privilege—especially after leaving a nation ripped apart by war. Some of the schools in Enugu were left in literal ruins from the bombings, and the ones still standing didn't have enough teachers or textbooks. So the fact that she was there at all, sitting in a classroom with four walls, electricity, desks, and chairs made her deeply grateful.

But as time went on, she became frustrated with how much more welcoming London seemed in those colonial propaganda videos than it actually was in real life. British newspapers talked about the sudden infestation of immigrants and what to do with *them*. She couldn't understand why everyone seemed to make such a big fuss over skin color. Since all Nigerians were Black, conversations about race were virtually nonexistent back home.

Many nights she'd come home to Arinze feeling like they'd made a mistake by moving there. She'd cuddle up next to him and tick through everything about England that irritated her. Things were so expensive. Her wages at the hospital would have covered a nice car back home, but here she could barely afford the bus fare. And making friends required so much effort. She'd smile at people in the street, and they'd stare blankly. Then there was the all-too-casual racism. "Keep Britain white" was a familiar refrain among far-right leaders. Nigeria may have had its problems but at least her Blackness didn't make her feel subordinate the way it did here.

Whether she liked it or not, England was her home now. Her husband was on his way to becoming a doctor, they had a flat, and they were expecting their first child. They'd sacrificed so much to be there. Arinze, who loved to stand tall in his three-piece suit, had taken a job cleaning offices to save £1 a day for Obiajulu's airfare to join him overseas. As much as the place frustrated her, she couldn't exactly pack up and go back to Oghe.

And so, one night she devised a secret plan. After preparing Arinze a plate of leftover rice and beans, she strapped on a thick coat and said something about retrieving a forgotten textbook from school. Arinze nodded without looking up from the novel he was buried in.

She slipped out into the night, her cheeks burning hot with shame even against the cool autumn air. She didn't like lying to her husband. But there she went, scooting along the quiet sidewalks like a secret agent on a mission until she came to a busy high street. The Odeon movie theater was all lit up as she imagined it would be, a dozen or so people lined up outside. The glittering marquee confirmed she was in the right place.

She'd heard *Live and Let Die* was violent and vulgar. The people in her biology class seemed to think violent and vulgar was a good thing, though, and that was the point. She'd overheard enough snippets of excited conversations in class to know this film was something they found interesting. Maybe if she watched it, and could talk about it with humor and enthusiasm, they'd think she was interesting, too. At least she'd be able to have a real conversation with them.

Obiajulu studied the other patrons carefully before stepping into the theater clutching her ticket. She had never been inside such a place.

The screen at the head of the large, dark room was massive. She wasn't sure where she was supposed to sit, and eventually settled in an aisle seat near the middle of the theater, which seemed to be the most popular spot.

She was excited at first when the screen came alive, but she had a hard time following the action. The characters spoke too quickly for her to understand everything, and the plot was confusing. She retrieved a pen and small notepad from her bag and began recording observations she thought might work best in a conversation with her teenage classmates.

—*Roger Moore, handsome.*

—*New York City—Black people, big buildings, drugs, guns.*

—*Fortune-teller, tarot cards, scary. Voodoo!*

—*Crocodiles, great escape.*

She was overwhelmed and exhausted by the end of the film. She thought it was strange that British people found that degree of bloodshed so entertaining. But as she navigated the dark streets back to their small apartment, she tried to replay the most significant scenes over and over in her mind, planning out exactly what she'd say in class. She was still humming the theme music as she tiptoed into the bedroom with a renewed sense of confidence. Arinze awoke when she snuck into bed.

"What took so long?"

"Um, train delay."

He went back to sleep instantly. Obiajulu lay awake deep into the early hours of the next morning, hopeful and excited about her new plan.

That next Monday morning, she marched into class fifteen minutes early with her head held high, ready with practiced lines to connect with her classmates at last. She waited until she got to her seat and caught the attention of the dark-haired girl to her right. Her heart was racing. But she had rehearsed this.

"How's exam prep going?" she asked with her best London accent, which wasn't good.

"Not bad. You?"

Technically, the girl *had* responded, even if she didn't look up.

"It's going well. I tried to do some studying this weekend, but I ended up going to the cinema."

"Oh?"

The girl was looking past Obiajulu, smiling at another friend.

"Have you seen *Live and Let Die*? Roger Moore was so captivating. I couldn't believe it when he escaped by skipping across the backs of those crocodiles. I was so frightened."

She waited for her new friend to respond, but nothing came. The girl was making notes in her workbook. Obiajulu had several more lines of commentary prepared, so she went on.

"And I didn't realize the dangers of New York City. All those drugs and guns! My goodness."

"Right."

"I loved it when Roger Moore rescued that woman. I just knew that they would fall in—"

"I think we're about to start." The girl nodded at the teacher, who had just entered the room.

My mother turned away and squeezed her eyes closed. This girl did not care about Roger Moore, about MI6, about car chases or gunfights. Or maybe she did, she just didn't care about discussing it with the Black, pregnant girl with the funny accent.

There are moments in life that seem small when they happen but are never forgotten. This was one of those moments for my mother. It was a soul-crushing defeat that even today—fifty years later—she can recall in exact detail. She realized she was painfully different from the kids around her, that she would never be one of them, and that nothing she could say or do would change that.

So she made a decision, not consciously perhaps, but a decision nonetheless, that she would press on with her studies and with life in London, with or without the friendship of her adopted countrymen. But she would do everything to ensure that her children had it better.

ꙮ

And that is why, after only a few months back home in London from my two-year training in Nigeria, she told me I was going away again. I was standing in the kitchen, marveling at the ease with which I could rinse my cereal bowl with warm water from the faucet, when she summoned me. She sat me down on the couch, wrapped one arm around my shoulder, and explained that I was to begin the next semester at St. Mary's, a Catholic school near Stonehenge, about one hundred miles southwest of London. After nearly two years of me living in the oppressive heat of Nigeria where I had to walk miles to fetch water, shave my head, and sit on broken toilets that did not flush, my mother was suddenly sending me to a posh boarding school in the English countryside, one of the whitest places imaginable. If towns could be opposites, a rural village near the English Channel was truly Enugu's upside down, and that's exactly what she had in mind.

I thought it was cruel at the time. On some level, it may have been. But I now know that her decision to send me away again came from a place of deep love—and deep fear. The soul-crushing pain my mother felt in that high school classroom decades earlier had followed her into adulthood and here to South London. Almost every day she was reminded that she did not belong still, that she was alone. That feeling was magnified, of course, after my dad died, especially when his network of friends and extended family eventually stopped coming by. I didn't know it then but my mother was terrified that I might grow up with that same sense of isolation.

Regardless of her reasoning, I was crushed. I was barely twelve years old, eager to reestablish myself in London after a challenging two-year stretch on another continent. I knew boarding schools were common here, but the last thing I wanted was to go away again. My mother sympathized, but sympathy did not cloud her plans for me. She was convinced that contrary experiences, whether I liked them or not, would ultimately teach me to navigate any and all environments I ended up in later in life, something she never quite mastered when she arrived in England. She believed there was value in oscillating between extremes, in being exposed to different surroundings, in being able to relate to people from all walks of life.

In Nigeria, children are naturally shuttled between worlds. Families who live in large cities like Lagos or Abuja, melting pots of various tribes and ethnic groups, are expected to speak a little bit of the three main languages in addition to English. I've seen Igbos speaking Hausa or Yoruba as though it were their native language. And those who enjoy the modern comforts of urban life often trek back to their families' villages in the bush every few months, where they're expected to live just as comfortably without running water or electricity. This degree of adaptability literally saved my grandfather from ethnic cleansing when the pogroms began.

My mother wanted to make sure as I entered my teenage years I would, unlike her, be able to navigate conversations, interviews, arguments, or interactions with people from any culture, race, or background—something my father could do surprisingly well. As far as she was concerned, the more exposure I had to people, places, and even schools that were different—and maybe even uncomfortable— the better off I'd be.

She didn't explain all this at the time, of course. Sometimes I wish she had. But as you know by now, that's just not how communication works with Nigerian parents and children. She told me only that it was for my own good.

I cried, knowing I had no choice.

🖎

Like at my elementary school in Enugu, the girls at St. Mary's wore knee-length uniforms and the administrators ruled with unforgiving discipline. But that's where all similarities ended. St. Mary's majestic stone façade with its Gothic spires, lush grounds, and virtually all-white student body was infinitely at odds with the plain concrete classrooms, dusty grounds, and barren fields that had defined my Nigerian school experience. And because my school in Enugu didn't have the money for cleaning staff, all of the students were required to pitch in. We cleaned the muddy mess ourselves during the rainy season. In the dry season, we swept the yard and helped cut the grass. In all seasons we scrubbed the toilets.

As my mother and I turned down St. Mary's stately tree-lined drive, I counted no fewer than four groundskeepers working on the perfectly manicured lawns. Part of me thought it might be a dream, that at any moment my grandmother would nudge me awake and I'd be back in Enugu about to draw her a bath. But I was wide-awake and intrigued by my new surroundings. At the very least, I felt certain I wouldn't have to scrub toilets at a place like this.

We parked and my mother guided me through the large arched front door and down the hallway. The classrooms we passed were stacked with luxuries I had almost forgotten existed: paintings on the walls, papier-mâché displays, and shiny new textbooks.

The head of house greeted us and politely informed my mother she was free to leave. I looked up at her with sad eyes but there were no tears. I was not afraid to be away from her anymore. Enugu had taken care of that.

"You'll like it here," she tried to assure me. "I'll call you once you're settled."

And just like that I was alone again, feeling anxious and afraid, just as she must have felt on her first day at that Walthamstow high school. But my fear did not last as long.

I stayed up late with my new dormmates that first night, trading stories about our mismatched lives. I told them about the world I had just left in Nigeria—the Sanitation Days and the trips to the stream and Mrs. Okonkwo's cane. We were all out of our comfort zones, and that was precisely my mother's intention. They peppered me with questions about life in Enugu, my family, and my hair, which had barely begun to grow back.

Even if misplaced, the attention felt good. I allowed them into my world, and I tried to cast the best light I could on Nigeria, a place they'd only ever seen in television news reports, if at all. We bonded in the coming weeks at the school dances and played rounders in the fields on warm weekends.

They really seemed to like the things that made us different. One evening a group of girls knocked on my open dorm room door, drawn in by the Nigerian beats playing on my stereo as I studied. They asked about the lyrics, which were in Igbo. I told them what I knew—most of our songs were about sharing gratitude with God. And with that, I jumped up from my desk and started shimmying across the room on my tip toes to the beat, shrugging my shoulders as I bounced up and down.

"This is how we do it back home," I said with a big smile.

Before long, an entire group of posh English girls was dancing to Nigerian highlife music. My grandmother would have fainted.

At first, I was shaky in the subjects like French and Latin that weren't taught at my primary school in Enugu, but I plowed on regardless. Eventually my grades began to soar. The commitment to learning what my mother had modeled in those late-night study sessions, backed by the intense discipline of Nigeria, followed me to this school. And as one of the most enthusiastic and hardworking students, I was made class prefect for both geography and English. The abrupt move to St. Mary's was working. Everything had fallen into place.

Then one afternoon, in an instant, it all came crashing down.

We were doing homework in a classroom after school one day, as we often did, when the head teacher, Sister Gilbert, stormed in and said she had an announcement to make. I scanned my classmates curiously.

"One of the girls in Year Eight has reported that her money box has gone missing. The person responsible has twenty-four hours to come forward. Otherwise, decisive action will be taken. If anyone in this room knows anything about it, please see me."

The sister paused for a moment, scanning the room, to ensure the gravity of her warning had properly sunk in before marching out.

A chorus of excited whispers filled the classroom.

"Who would do something like that?" "How much do you think was stolen?" "I wonder if my things are safe."

In an all-girls boarding school in the English countryside where nothing particularly exciting happened, a whodunit mystery was as good as it got. By the next afternoon, no one had come forward. Sister Gilbert returned, and this time with the head of house.

"Since the person or people involved have not been willing to con-fess, we will search every one of your rooms to get to the root of the matter. We will not tolerate theft at St. Mary's. Sister Elizabeth and I will take each of you to your dorms one at a time to conduct the searches."

The girls looked at one another with wide eyes, not sure whether to be frightened or delighted by this unexpected drama. It was becoming something from a Nancy Drew novel. Sister Gilbert eyed the classroom and pointed to a girl sitting in the front row.

"We'll start with you. The rest of you will stay seated until you are called."

The first girl returned quickly, and the sister called on the next one, and the next. It went like this for more than an hour until my turn came. Sister Gilbert beckoned me over and led me upstairs to my room. I stood silently as the nuns searched under my bed and rummaged through my drawers, examining the various trinkets I had on my dresser: a beaded necklace from Enugu, a blue-and-white friendship bracelet, and a small colorful change purse.

You may know where this is going, but I did not. I was actually annoyed it was taking so long—I had homework to tend to—when I heard Sister Elizabeth gasp.

"What is this?"

She turned toward me slowly, and in her hands was a thin red metal box. I was stunned. I had never seen it before, but I knew what it was. We all knew what it was.

"That—I-I-I didn't put that there. Someone else must have—I didn't do it!"

Sister Gilbert slapped away my protests with an angry wave and without a word grabbed me by the shoulder and yanked me out into the hallway and back down toward the classroom. We walked in together, her still holding the money box as she announced that the culprit had been found.

The room was dead silent as I skulked to my seat, my head down, afraid to meet the judgmental stares that needled my flesh.

"I didn't take anything; someone else must have put it in my drawer," I whimpered as I sat down.

I stared at the open notebook on the desk in front of me, trying to come up with a solution. My mother would believe me; she had to believe me. But in the eyes of my teachers—and probably most of my friends—I was the villain. The mystery was solved, and I—the only Black child in my year—was a thief.

I'd been at St. Mary's for only a couple of months, but my life there was transformed in an instant.

Girls at eleven and twelve can be wonderfully sweet and loyal. They can also be cruel. The girls of St. Mary's turned their backs on me almost immediately. A few pretended to take my side at dinner that night, but within a few days I'd been completely ostracized. The teachers moved me to my own isolated room on the other side of the school, far away from the other children. My class prefect titles were stripped and given to other students. Essentially friendless, I began to spend weekends wandering alone up and down the long driveway to pass the time between lunch and supper. I was devastated, and so was my mother, who never doubted my innocence.

By the end of the week, my mother had arranged a meeting with the headmistress. When the time came, she abandoned any sense of British etiquette she usually tried to maintain around white people. There she was, in a room full of English nuns, flapping her arms wildly, pointing and shouting at the top of her voice, her refined English cadence replaced by a thick Nigerian accent. There was too much at stake to be self-conscious about her village ways at a time like this.

The headmistress was unmoved. She calmly pointed to the evidence: I'd been caught red-handed. The money box was in my drawer. There were no other suspects. It was an open and closed case. And it was grounds for expulsion.

My mother was not going to let these nuns ruin her daughter's academic career for something she didn't do, and she couldn't bear the thought of another expelled child. She wrote letters to the school and came in for more meetings with the nuns, who in the end agreed to let me stay at St. Mary's. I would have regular check-ins with Sister Elizabeth to make sure I was on the right track. If there were any more issues, I would be sent home.

My mother had technically won by preventing my expulsion, I suppose, but we both knew it was a hollow victory. In the weeks that followed I remained an outcast among my peers, distrusted by my teachers, and consumed by loneliness. Most nights I would stand in line for an hour to get a few minutes on the dorm's only pay phone, just to hear my mother's voice.

As much as she wanted to fix things, there was no quick remedy. Even if she removed me from St. Mary's in the middle of the semester, it would take several weeks to line up another school. And she knew that most good schools would have waiting lists at least until the next semester, if not longer.

With few satisfying options, she opened up for the first time about her own struggles in England, feeling like she was invisible and alone no matter how hard she tried. At first, she was willing to do almost anything to find some sense of acceptance from her new community. She changed how she spoke, what she ate, how she had fun. Nothing worked. At some point, she began to accept herself as she was, even when none of her neighbors would. She wore African clothes to our school functions—her orange abada wrapper and blouse were among her favorites—knowing that people would stare, and she spoke Igbo in public whenever she had the chance. In this way, she found some semblance of peace.

It was what I needed to hear, perhaps, but it certainly didn't make my days any easier. In Enugu, I didn't fit in because I was Oyibo. Here, in the country where I was born, I was still an outsider.

Maybe it was all part of my mother's plan. Thinking back now, it's obvious that shifting me to and from wildly different environments through my formative years would create problems—problems I would inevitably be forced to overcome and learn from. But in that fall of 1995, as a twelve-year-old girl just starting to feel grounded again, the last thing I wanted was more upheaval.

I kept hoping and praying that somehow I would be redeemed, that if I continued proclaiming my innocence then surely they'd realize I was telling the truth. It never happened. Most students were embarrassed to be caught talking to me, socializing with a thief. Teachers felt it was their duty to ignore me or isolate me to make me change my behavior even though I hadn't done anything wrong. I was broken, but I refused to let them win. I kept going to class every day. I kept preparing as my mother taught me. I kept going to the dances. I kept trying to navigate this new environment she'd immersed me in.

And one day, just before the Christmas break, Sister Gilbert called me into her office. I sat down anxiously, wondering what sort of sinister plot had befallen me this time. My mouth fell open when she apologized.

Another student had come forward to confess. *She* was the one responsible for stealing the money box and hiding it in my drawer. After watching me get bullied and shunned for several weeks, she apparently came clean to an older student who'd forced her to turn herself in.

I was redeemed. But I was still angry. The target may have been lifted from my back, but I'd been forced to endure a torturous existence for almost the entire semester for something I did not do. Before I left her office, Sister Gilbert offered a parting thought:

"There are no mistakes in God's world."

As I walked around the school grounds that afternoon, part of me was relieved that I had been cleared of the false accusations, but in reality, I was crushed. Another girl—or girls—had intentionally set out to frame me. It was probably someone I had laughed with or danced with, someone I thought was a friend.

There was no public apology, no formal acknowledgment of my redemption, but word soon spread among the students that I was innocent, and the teachers who'd stripped me of my prefect titles were suddenly friendly again. But they never apologized and the titles stayed with their new owners.

I returned home for Christmas that following weekend. After just one semester at St. Mary's, I never went back. Despite the school's apology, my mother refused to let it go. Some sins cannot be forgiven. She enrolled me at a small girls' school in South London for the following semester. I would spend the rest of my high school years there, and I thrived.

Sometimes it feels like St. Mary's was my Walthamstow Forest, a sting that will never completely go away. But no one among us escapes our formative years unscathed. My mother understood that better than any of us. There is always betrayal, humiliation, injustice, but I now know that I am a stronger person because of the totality of my vast and varied experiences—the changing schools, the shifting environments, the new faces.

When I was barely a teenager, I had no real appreciation for the power of my mother's lessons. Before the end of high school, however, I would become a believer.

Chapter 10

People like me do not go to Oxford University.

I am a first-generation African immigrant, raised by a single mother who struggled to keep the heat on in a gritty home in East London before we moved south of the river. Oxford is a place of kings, literally. It is a 900-year-old symbol of privilege, class, and status that has educated more than a dozen British prime ministers and the heads of state from at least twice as many other countries. It is consistently ranked as one of the world's best universities, but in the United Kingdom, it is so much more than that. Oxford is a key that opens almost all doors and offers the real possibility of generational success. Known for its ridiculously difficult application process, the vast majority of good students don't even apply.

I didn't know any of that the first time my mother dragged me there. It was the late spring of 1996, and I was just beginning to settle into the social scene at my new school in South London. And at only thirteen, getting into an elite university—or any university—was not a priority, especially with at least another five years of secondary school left.

But Nigerian mothers are nothing if not planners.

It was a cold Sunday morning in February when my mother forced me into the car for that ninety-minute trek west of London. We didn't know anyone at Oxford, and as a kid, I obviously didn't have an appointment with an admissions counselor. We hadn't even signed up for a tour. My mother just wanted me to be there—to see and feel and smell the institution that had educated so many of the world's decision makers. She wanted to show me what was possible. She wanted it to seem real, to seem reachable—for both of us.

On paper, Oxford was an impossible dream. The university required years of proven academic success, was a magnet for people of privilege, and featured a student body that was only 2 percent Black. All of that should have been a deterrent. It wasn't.

My mother was used to things being hard. Even with the war behind her, she was forced to endure the never-ending pressure of being a single mother, the constant threat of violent crime at the pharmacy, and the daily reminder of her otherness. Oxford was a dream so big, it could somehow justify the struggle.

She cranked down her window and inhaled deeply as we entered Oxford's city limits. She seemed to be talking more to herself than to me as she ticked off the school's extraordinary alumni, a list that included Colonel Ojukwu, the Nigerian military leader who led Biafra during the civil war. There were others too: Lewis Carroll, Stephen Hawking, Margaret Thatcher—even Albert Einstein lectured here! The nuns from her Nigerian schools had referred to Oxford with such reverence that she'd always envisioned it as a magical kingdom of sorts on par with Narnia or Oz.

Her eyes widened as we stepped out onto St. Giles', one of the main thoroughfares that fed into the city center. Here she was, walking where legends were groomed. Over the course of our day together, we learned Oxford didn't really have a central campus. Instead, it was spread out across a series of centuries-old buildings in the center of the city, scattered among local shops, terraced homes, and cobbled streets. And there seemed to be more bicycles than cars. The university was also made up of a collection of thirty individual undergraduate colleges, each with its own dining hall, libraries, and dorm rooms that housed roughly six hundred students.

We walked down the bumpy stone streets from one castle-like structure to the next, my mother stopping to examine the significance of almost every building and college with the help of her guidebook. She read aloud in her thick Nigerian accent:

"The New College. Founded in 1379. Oh my goodness. It says here there are tunnels under the Bodleian Library. Tunnels!"

"This looks like a palace, but they call it Christ Church College. Do you know how many British prime ministers went here? Look, it says fourteen. Fourteen prime ministers studied right here."

She was constantly nudging me and pointing as we passed students, as though we'd stumbled across an endangered species on safari.

"That could be you someday."

I was far more focused on my freezing fingers and toes than the history of one of the world's oldest institutions of higher learning, but as my mother and I wandered around the storied campus for hours I could not help but be moved by her excitement. I began hounding her with questions befitting of a thirteen-year-old: "How could anywhere be a thousand years old?" "Why is everyone riding bicycles?"

My mother was beaming when we finally returned to the car. It was a smile that didn't fade during the long drive home, or even that night as she scrambled to cook dinner for Kandi and me before talking through our weekly reading assignments and helping me study. She should have been bone tired—I know I was—but long after we were all asleep she was still awake, flipping through the Oxford guidebook, rereading the history of each of the colleges, their various personalities, and their most influential alumni.

The girl from Enugu could hardly believe what she had seen that day. But somehow, it felt achievable. Obiajulu had managed to earn the highest academic honors after living through unspeakable bloodshed; she had passed her pharmacy exams while working part-time and raising two young children. If she could do that, then her daughter could certainly do this.

I didn't hear much about Oxford in those subsequent months as I settled into a healthy routine at school. There was a heavy focus on academics, and that was something I'd been taught to take very seriously. But I began to drift a bit after I turned fourteen, succumbing to the usual set of teenage distractions—boys, parties, staying up late. Like any teenager, I craved freedom, and with such a strict mother, I needed the relief that temporary rebellion offered. I was beginning to realize that London presented a whole host of nightlife options for teenage girls, and if I was smart about it, I could partake without my mother knowing.

One night I crept into the house after a late night out, catching a creaky floorboard as I tiptoed past her bedroom. Within seconds, the door was open and she was on fire.

"Where have you been?"

I stood rooted to the spot like a frightened deer, not sure whether to lie or tell the truth.

"Don't make me ask you again."

"Just a friend's house."

"Until 3 a.m.?"

I shrugged my shoulders.

She was angry. But rather than punishing me as she ordinarily might have, she took me back to visit Oxford. Less than a year after our first trip, we made the sixty-mile pilgrimage once again to wander those hallowed streets. This time we stayed a little longer. And again, we returned happy and inspired. In this way, a mother-daughter tradition was born that spanned the entirety of my teenage years. Whenever I misbehaved or stayed out later than I should have, I was never grounded or spanked or forced to do chores. Instead, my mother would take me back to visit Oxford, to show me something better to aspire to.

In the early days, I never particularly wanted to go, and she certainly didn't have the time to spare, but somehow we always found a way. It was a ritual that I grew to love. Those trips carried us beyond the struggle of our daily lives, for an afternoon, at least.

Before the age of eighteen, I'd been to the university half a dozen times. I knew the names of the buildings and the colleges I liked best. I knew how the students dressed, how they got around town, and where they ate. Going to Oxford stopped being a dream. It actually felt possible.

☙

I was sixteen when my school held a parent-teacher meeting to discuss each student's university prospects for the first time. After so many uncomfortable interactions with our teachers, this was the one meeting my mother actually looked forward to. If nothing else, she was well prepared. All the weekly reading lists, the late-night study sessions, time with uplifters, and the trips to Oxford had been building to this moment. The teachers would see, as she had, there was only one place her daughter should go to university. She had done anything and everything she could think of to make it so.

What my mother did not know, could not have known, was the extent to which my teachers controlled my academic future. University applications required teachers to post *predicted grades* for our A level exams—and Oxford required three A's just to be considered. (Only about one in ten students in the whole of England typically get three A's.) Beyond convincing my teachers to give me the highest predicted grades, I would also need them to stake their professional reputations by writing personal letters of recommendation declaring I was Oxford material.

This was asking a lot of the teaching staff. I was a good student, but nowhere near the genius status expected of the Oxford bound. And we were all aware that very few Oxford students were Black. All signs suggested it was a long shot.

My mother stepped into the large classroom with sky-high expectations, ready to receive professional validation for what she'd spent much of the past year planning. Mrs. Davies greeted her warmly, and once they were both seated, she quickly turned the conversation to my future.

"Your daughter has been one of our stronger students. If she continues to work hard, I see a real possibility she could attend the Universities of Leeds, Durham, or Edinburgh."

My mother's grin dissolved instantly. Those may have been very good universities, but coming from Africa, she hadn't heard of any of them.

"Um, excuse me, Mrs. Davies, that is a good list, indeed. Thank you very much. But I was hoping she might also apply to Oxford?"

After a few moments of silence, the silver-haired teacher cleared her throat uncomfortably and repeated my mother's question.

"Oxford?" It was as if she'd been forced to say *Voldemort*. Her brows furrowed with incredulity; she pursed her lips and looked to the heavens before speaking again.

"A university like Oxford is far too competitive. That would almost certainly be a waste of everyone's time—the school's and mine. Oxford needs years of demonstrated academic excellence. Your daughter has done well in school, but Oxford requires a higher level."

A lifetime of fighting had prepared my mother to push back. She had come too far and sacrificed too much to accept this teacher's dismissal.

"We've been working toward Oxford. We've visited the university lots of times. I've done my research and spoken to a lot of people. It's not that far-fetched."

Mrs. Davies did not hesitate this time.

"I respectfully disagree. The universities I mentioned are excellent options. I'd encourage you to prepare her for those, or possibly Imperial if she really excels. Let's not fill her head with silliness."

Realizing she was wasting her time, my mother stood up, thanked the teacher coolly, and walked out of the room.

She drove home in furious silence. She knew that things were different in England, that people in power had their own way of doing things that preserved a certain order that went back centuries. But she didn't expect those people to actively suffocate her daughter's ambitions, especially given how hard she was trying.

In Nigeria, communities exist to lift up the individual. There are always exceptions, but there is generally a shared belief that personal success ultimately reflects well upon the entire community—that the success of any one person makes every person stronger. In that way, Nigerians have a common interest—a duty, almost—to help the people in their community reach their greatest potential.

In England, as my mother was learning, there was a different set of rules. She couldn't just rock up and force Mrs. Davies to endorse our dreams. She couldn't yell, haggle, or cajole as she might have done back home in the village. She couldn't even force her to be open to the possibility. Even if it was within my reach, she was operating in a system in which the power brokers were usually too concerned with maintaining structure and order to roll the dice and take a risk.

My mother marched upstairs when she got home, flung open my door, and told me what had happened.

"They don't think you can apply to a place like Oxford. It's too competitive, apparently."

She let out a deep sigh and plopped down on the edge of my bed. I was disappointed. I had started to believe Oxford was actually within my grasp, especially after seeing a fellow Nigerian, John Chinegwundoh, graduate at the top of his class from Oxford's rival, Cambridge. At the same time, I knew that I was a good student—not a great one—and I wasn't entirely convinced that my teachers were wrong.

My mother wasn't having it.

"I'm going to figure out a plan."

That evening, she spent what felt like hours pacing her bedroom like a caged lion, hungry to overcome the feeling of powerlessness the parent-teacher meeting had created. She needed to prove them wrong, to prove to everyone in this country that her family, and by extension the village of Oghe, could achieve something magical. She was convinced that Oxford was the ticket to a better life, and God help anyone who stood in her way.

Despite her passion, the options were somewhat limited. She could no longer spend hours each night teaching me in advance. I was a teenager now and the material—geometry, algebra—was too difficult for her as I grew older. She couldn't afford tutors either. If Oxford required a higher level of academic excellence, she was going to have to get creative.

One evening after dinner, my mother headed to the kitchen to clean up. As she soaked the plates in soapy water and scrubbed away specks of plantain, she suddenly heard a loud noise coming from the living room. It sounded like I was screaming in pain. Worried, she dropped the plate in the sink, bolted to the hallway, and burst into the living room to see if I was okay. The look of panic on her face turned to dismay when she realized I wasn't crying but laughing. My sister and I were in the middle of our nightly ritual, standing in front of the television reciting every single commercial that appeared on screen. We knew every jingle, every sound bite, every facial expression by heart. I could regurgitate most of the commercials with my eyes closed, and my sister, the real comedic genius, could even act them out. My mother looked on, shaking her head in disappointment, horrified that her children spent so much time memorizing such useless information.

It was after 10 p.m. when she entered my room with a stern look.

"I have a solution about getting you into Oxford. It's simple. It won't cost a penny, and I'm almost certain it'll work."

She wasn't going to stop me from seeing my friends or lay down some draconian curfew. There was one obvious distraction that loomed above all others and represented a clear and present danger to my meandering teenage attention span: TV.

She told me that from that moment on—just under two years before I would even start university—I was barred from watching any television whatsoever until I had an actual Oxford acceptance letter in hand.

I thought she was joking at first. Who had ever heard of such a thing—no TV? But I realized she meant business when she walked over to the beat-up 13-inch television balancing on my nightstand.

"That will have to go. And I do not want to see you anywhere near the TV in the living room."

I sat in shock as I watched her scoop up the small set and leave the room. There was no more discussion.

It took me a while to process what she had done. With both of my older brothers already out of the house, television was my primary form of entertainment—and my most important connection to pop culture, which dominated the high school social scene. My little sister and I spent almost every day after school and most weekends entranced by the best of mid-'90s British and American television. We were obsessed with *The Queen's Nose*, *Eastenders*, and *Moesha*. Conversations at school revolved around the latest MTV videos or the latest scandals in *Byker Grove*. At sixteen, TV was a huge part of my life.

With some time to formulate a more thoughtful defense, I tried to plead my case the next morning at breakfast. I explained that I'd be a social outcast without TV. More importantly, I'd miss out on educational programs. Didn't she want me to keep up with current affairs on the news?! And what if Chiwetel got a role on a big TV show or movie? Wouldn't I be able to watch it? She held firm. Beyond the occasional nightly news report, recreational television was out of the question. As far as she was concerned, television was stealing from my future.

But as a single mother who ran a small business six days a week, she couldn't monitor my television intake around the clock. She warned of consequences if I disobeyed, but the rewards outweighed the risks at first. As soon as she'd leave for the pharmacy on Saturday mornings, Kandi and I would race to the living room and spend hours watching music videos or *The Simpsons* reruns. It was like that most days after school until we heard our mother's Honda growl in the driveway.

It took less than a week for her to catch me in the act. She walked in on me while I was engrossed in a particularly intense episode of *Quantum Leap*.

"This won't be happening again," she said somewhat ominously, before ordering me to my room.

The next day after school, my sister and I assumed our normal positions in the living room, and I flicked on the TV. Nothing happened. I pressed the button again. And again. Nothing. I peeked around to the back of the television, and to my horror, I discovered the cause of the problem.

The power cord was lying there in two pieces.

My mother had literally taken a pair of scissors and cut the cable in half.

I gasped. *This woman has snapped.*

I knew she was capable of a lot but I never anticipated she'd intentionally destroy a major electronic appliance simply to boost my grades.

I didn't realize she was operating under a different code, the code of the village, in which long-term success—even survival—often required serious short-term pain. During the war, she'd try to save half her daily food portion for the next day, no matter how small it was. That sort of thinking, that commitment to delayed gratification and extreme discipline, had saved her life as a child and seeped into every aspect of her parenting style. She was more than willing to ask us to live without heat, sufficient sleep, or television if that was required to live a better life someday.

I screamed and cried that night at dinner. I called her callous and mean. I begged her to reconsider.

"One day you'll thank me."

Hard work means different things to different people. Until then, I'd always considered myself a hard worker. Sure, I liked to zone out in front of the TV sometimes, but I also spent hours every night studying. I was engaged in all my classes, and at my mother's urging, I requested practice tests and special assignments during school vacations. Maybe I wasn't the best student at school, but I was much closer to the top than the bottom. I excelled in languages, held my own in mathematics, and got a few A's and mostly B's in practice tests. That was good, just not good enough for Oxford.

I wasn't ready for life without television at first. I sought out other distractions right away—anything to avoid having to spend every spare moment studying. The phone became my new best friend.

There was a landline just outside my bedroom, and I usually carried it into my room after school, stretching the cord under my bedroom door for privacy. I had my close circle of friends, but really, I'd talk to anyone who'd listen—boys, neighbors, distant relatives, it didn't matter. I'd spend hours sitting on the floor propped up against the door with the receiver to my ear. With no TV, it was my only connection to the outside world. I needed to know what was going on.

My mother noticed right away that I seemed to be replacing one diversion with another. At first, she tried verbal warnings. When that didn't work, she started yanking the cord out of the wall socket when she thought I'd been on long enough. I don't remember how many times I was in mid-conversation when the phone line suddenly went dead. The telephone tug-of-war quickly became a nightly ritual—until she found a creative solution for that, too.

One Sunday morning she returned from running errands carrying a large box. Without a word, she hauled it to the upstairs landing and unplugged the phone from the wall.

"We won't be needing this anymore."

She opened the packaging and pulled out an unusual-looking rectangular contraption. Gray and boxy, it looked like a cash register built into an old phone. I looked more closely and noticed a narrow gap on one side.

"What's that, Mum?"

"A pay phone."

"A what?"

"A residential pay phone. If you want to talk on the phone from now on, you'll have to pay for it yourself."

She calmly unraveled the cord and jimmied it into the wall socket and stuffed the old phone into the empty box.

"It takes coins. If it's an emergency, you can ask to reverse the charges."

My mother had somehow found a residential pay phone to put just outside my bedroom. I was stunned. I'd seen a handful of these bulky devices in office buildings and waiting rooms around London, but I'd never seen one in a house. Starting that day, if I wanted to talk to my friends—or anyone—it would cost me 20 pence a minute.

At sixteen, I barely had any money. I didn't have a formal allowance. My main source of income was the change from my bus fare to school and whatever I could find in the couch cushions. On a good day, ten minutes was all I could afford. So I began to use the phone if, and only if, what I had to say was urgent. My conversations with friends became incredibly concise.

One appliance at a time, my mother was essentially re-creating the conditions that led to her academic success during her turbulent childhood. There was no electricity—never mind televisions or telephones—back in Oghe. There were no boys to talk to during the war, no matter how much she would have loved to spend time with my dad. My mother's high school years were all about survival and study. When she wasn't dodging bombs or searching for food, she was reading donated books and sitting through her mother's impromptu classes in the bush. They didn't always have textbooks, but somehow Mama Nnukwu doled out plenty of homework. That's just how it was. And that's why my mother graduated at the top of her class when most people would have been happy to graduate at all.

My mother had worked hard because she wasn't just working for herself. She was working for *her* mother, a woman who had sacrificed meals to give her a better life; she was working for her village, which expected great things from all its children. Her successes and failures were their successes and failures, just as mine would be.

Installing a pay phone in our house was a drastic move, but like chopping the television cord, it was effective. I would come home from school each day with a simple choice: I could sit around staring out the window for hours or I could study. I got home most afternoons around

4 p.m. and didn't go to bed until at least 10 p.m., so I had a full six hours to fill. Reluctantly, I began to fill those hours at my bedroom desk studying and doing homework and studying some more. I also started reading around my subjects, diving deep into the complicated alliances that fueled the Spanish Civil War and the works of absurdist playwrights like Samuel Beckett and Eugène Ionesco to help fill my time.

Within a month or two, this chore of constant studying became part of my normal routine. I brushed my teeth. I ate lunch. And I studied at my desk. Within a few more months, the routine became something I reluctantly began to enjoy. My grades at school improved right away. Not only was I getting straight A's, but in most of my tests I was getting every answer correct.

One day before school, my French teacher tracked down my mother in the parking lot to ask about my sudden surge. When my mother explained that it was simply about eliminating distractions, my teacher asked for advice about what to do with her own children.

Before long, everyone at school knew what was happening at home. I missed my old life, of course, but I couldn't ignore my dramatic improvement as a student. I went from being a teenager who could recite the lyrics to every pop song and dance along to every '90s music video to someone who actually had a handle on postwar British politics, monetary and fiscal policy, and nineteenth-century literary greats. I started a study group with friends who appreciated the fact that my home was basically a glorified library with bedrooms.

⟿

Nine months after the first meeting, my mother went in for another parent-teacher conference. She wasn't going to take no for an answer this time. With nearly perfect grades, there was no legitimate reason to block me from applying to Oxford. Mrs. Davies was still somewhat skeptical—"Oxford isn't for everyone," she warned my mother, who continued to press. But after a few minutes of debate, she finally relented and gave her blessing. Over the coming days, more teachers signed off. I would have the predicted grades I needed. I'd have the recommendations. Now I just needed to get in.

Oxford's admissions process is unique. Aside from the application, there are grilling interviews to contend with and a handful of specialized tests to sit. The tests are impossible to prepare for, as they are designed to test *how* you think, and not necessarily whether you get most answers right. My tests were to be held on a Tuesday in December.

When the day came, my mother actually closed the pharmacy early—to drive me those familiar sixty miles west of London. She later told me she spent much of the week praying alone in her bedroom as I sat in a classroom packed with some of the world's most ambitious students going through a series of complicated French and Spanish translations. The rumors were right. The tests were brutal.

Next came the in-person interviews. I had done enough research to know this was the most difficult part of the screening process—and the most difficult part to prepare for. It wasn't going to be a friendly get-to-know-you conversation with an alumnus or two as it sometimes is in the States. Prospective students face a panel of professors from their chosen area of study who are known to be merciless.

They asked things like, "What is language?" or "How would you work out how many golf balls can fit on an airplane?"

The interviews were designed to be unpredictable. The distinguished professors, or dons, as we called them, were trying to determine whether prospective students had the intellectual capacity to think independently, to be stretched, to go beyond simply regurgitating memorized information.

The interviews with our desired colleges were expected to be less than thirty minutes, but we were asked to be prepared to stay for several days in case more than one college was interested. My final interview was set for 11 a.m. on Friday morning at Keble College, Oxford University.

I barely slept the night before, my brain racing through dozens of possible questions and answers I had written out and rewritten and written again to prepare. Given that I had chosen to read French and Spanish, I knew they could ask me about poetry, literature, or even the politics or policies of the French or Spanish governments. If I was going to fail, it was not going to be for lack of preparation.

That morning, I wandered across the college quad through a trickle of anxious students. I arrived almost half an hour early and tried to look calm as I went over my memorized notes in a small seating area outside the interview room. Without warning, the silence was interrupted by the creak of the door as a well-dressed boy walked out, jokingly wiping the sweat off his brow.

I smiled momentarily and then heard my name. It was happening. I felt like I was walking back in time as I entered the room; its walls were covered with rich wood paneling and large portraits of some of the college's most prominent benefactors, wardens, and honorary fellows. There were two older professors sitting shoulder to shoulder in the center of the room. In front of them was a single chair. My chair.

One of the men welcomed me somewhat coldly. I sat down and he handed me a piece of paper. It was a complex nineteenth-century French poem by Charles Baudelaire. He asked me to translate it on the spot. I was expecting something like this, although the subtleties of a 150-year-old French poem made me stumble at first. I kept going, gaining confidence as I went.

The other professor interrupted before I'd finished. He wanted to know which rhetorical devices were used, asking a few questions about the metaphors, tone, and figurative language. He wanted some of my responses in French. I stuttered through some of my answers. We moved quickly into twentieth-century French literature. They wanted to hear about my favorite authors from the era, what specifically drew me to them. Again, I was prepared. I spoke of my interest in Albert Camus and existentialism and the significance of his writings, especially given the political context of the French Resistance, and later his oscillating stance on Algeria's independence from France.

I connected the concept to my family's struggle during both independence from Britain and the Biafra war, when the government failed to protect so many of my countrymen from ethnic cleansing and famine. Maybe it was a stretch, but I wanted them to know about the strength and resilience I had inherited from my Nigerian culture.

I was only in the room for twenty minutes when they thanked me for my time and asked me to leave. I stood up slowly and looked my panel in the eye. Two years of collect phone calls, no television, and studying around the clock had taken me to this moment. "It's been a pleasure."

My hands were shaking as I reached for the brass doorknob. I thought I had done well, but I was far from confident. I walked down the ornate hallway toward my dorm room in a trance. The stress of the interview was gone, replaced by a far heavier weight. The hope and pride of my mother, my family—even my village four thousand miles away—were at stake. There were no moral victories for coming this far. I had to do the impossible.

The next few weeks were torturous. At the time, I was seventeen and stuck at home with nothing to do but wait for the most important news of my life thus far. Two weeks passed and nothing came. When Christmas vacation arrived, I spent every morning in bed silently waiting for that day's post to fall through the mail slot. As soon as I heard that metal clang, I'd sprint downstairs in my pajamas and flip furiously through the mess of magazines and letters.

Early one Tuesday morning, I heard the mailman's steps just outside the door. I leaped out of bed and raced down the stairs. As I scrambled through the mail, my heart froze when I flipped to an envelope post-marked OX1.

This was it.

It was thin, so thin that it could only contain a single sheet of paper. A subtle feeling of dread crept into my belly as I tore into the letter. Every cell in my body holding its breath.

I was so nervous that my eyesight was blurry. I had to read the first sentence three times before I understood what was happening.

It is our pleasure to welcome you to . . .

I couldn't even read to the end before I started screaming and jumping and crying. I stretched my arms up toward the heavens, thanking whoever was up there. When I caught my breath, I turned back to the letter, my heart exploding, as I read and reread every word.

I ran upstairs to the hallway telephone and grabbed the receiver before realizing I didn't have any change to call my mother. I skipped to the closet where she kept her coats and rifled through the pockets, desperate to find the 20 pence I needed to share the life-changing news with the woman who made it possible. The pockets were empty. I would have to reverse the charges.

I gave the operator the number to my mother's pharmacy and listened excitedly as she agreed to pay for the call. I could tell by the tone of her voice that she was busy, likely with a long line of impatient customers.

"Mum, guess what?"

"What?"

"I got in."

"What?"

"I'm going to Oxford, Mum. I have the letter in my hands."

There was total silence for a few moments. She was almost whispering when she spoke again.

"I'm going to call you right back. Don't move."

It took her almost fifteen minutes to clear the line of customers and return the call. She told me to read every single word of the letter, instructing me to enunciate and speak loudly. I could hear her sniffling as I read, knowing that tears were trickling down her cheeks, just as they were mine. The girl from the village, who had escaped genocide, bomb blasts, and starvation, had a daughter going to Oxford.

The whole family came home for Christmas that weekend. My siblings had their own successes to celebrate. Obinze had started his own high-end fashion distribution company operating in London, Paris, and Milan. He specialized in positioning designers in big department stores like Harrods and Selfridges. Chiwetel had already gotten his big break with a role in Steven Spielberg's *Amistad* and filming would soon start for his break-out role in *Dirty Pretty Things*. But I was the star that day. My big brothers gathered around me, arms folded, marveling at what had happened.

"Oxford? You mean *the* Oxford?" Chiwetel joked.

"You're set for life, aren't you now?" Obinze added.

Obinze explained I wasn't set because of the doors Oxford would help open, although doors most certainly would open. It was because I had done the unimaginable. I had every excuse to fail: I'd lost my father; I had a mother who was badly overworked; I'd bounced around half a dozen schools; and still, I had done it.

"There's nothing you can't do now."

We called my grandparents that night on the phone my mother kept locked away in her bedroom. They were about to head to the village for the holiday with most of our extended relatives. My grandparents screamed and cried almost as loudly as I had when I told them. They'd been to England only once in their lives, to help us in those months after my father died, and they had certainly never seen Oxford. But growing up in colonial Nigeria, they knew what Oxford represented, its mythical status, and they knew what it meant for a girl like me to get in—what it meant for all of them.

I later learned that the news spread quickly across our community in Enugu. People in Oyofo-Oghe gathered in the village later that week. Some danced, cheered, honked horns, and drank deep into the night, long after I was asleep. They celebrated my achievement as if it was their achievement, because it was, whether they knew me or not. It was a triumph that validated generations of suffering and sacrifice in a nation whose people had every right to lose hope. But that Christmas, there was hope—real, verifiable hope written on a crinkled letter postmarked OX1 a continent away—for them, and for all of us, that proved anything was possible.

Chapter 11

Having lived in New York City for more than six years, I had no shortage of favorite places. The Port Authority Bus Terminal, a perpetually overcrowded labyrinth of escalators and grimy corridors in Hell's Kitchen, was not one of them. I had been invited to go out with friends at one of those trendy tapas places on the Lower East Side that night, and as much as I would have rather gone there, I was on a different kind of mission.

In my hand was a bus ticket to Providence, Rhode Island, a place I had never been to and only knew existed because a close friend moved there after grad school. I stumbled across my gate and boarded the Greyhound bus, squeezing down the overstuffed aisle toward a window seat near the back. I was tired. I reached into my backpack and pulled out a sweatshirt, which I scrunched into a ball and wedged between my neck and the cool glass. I was hoping to sleep, but with the excessively bright overhead lights and the cramped quarters, I wasn't optimistic.

We pulled out of the station and soon found Interstate 95. When sleep didn't come, my mind wandered across the six years in America that brought me to that point.

I'd left Oxford with degrees in French and Spanish and promptly bought a one-way ticket to New York to attend Columbia University's graduate school of journalism. About halfway through Oxford, I'd decided that I wanted to be a news reporter. It may have been an unlikely choice, given my mother's views on television, but I was genuinely attracted to a profession that demanded constant study, research, and curiosity. And the idea of being thrust into new environments from one day to the next thrilled me.

But five years after earning a master's degree in journalism, I wasn't exactly setting the world on fire. I'd spent more than three years answering phones at a production company before becoming a freelance television reporter for a small station in the Bronx. I was still hopeful about my future in journalism—too hopeful maybe, given the insanely competitive job market and my underwhelming résumé.

My thoughts turned to the week ahead in Providence. I knew that if everything my mother had taught me about generating success was true, this could be one of the most important trips of my life.

I arrived after midnight, bleary-eyed from the three-hour journey. Providence was a small and gritty city, a working-class town where crumbling mills sat abandoned along the highway and the Italian mob still clung to its last remnants of relevance. I heard someone shout my name after I stepped off the bus and was relieved to see my friend and host for the week, Nneka, standing outside her car.

I ran over and we hugged and screamed like schoolgirls. It had been years since we'd seen each other at Columbia, but like it is with some close friends, it felt as if no time had passed at all. Nneka was also the child of Nigerian immigrant parents—parents whose rigor for academic learning could have made my mother blush. As a result, Nneka and her siblings had the Ivy League well covered, with degrees from Princeton, Harvard, and Columbia.

Education aside, Nneka and I were cut from the same cloth. We both came from tough-love immigrant families with succeed-at-all-costs mentalities that guided our decisions well into adulthood. That's why she completely understood what I was doing in Providence.

After a quick drive to the city's West End, we crept up the staircase of a brick apartment building to Nneka's studio. It was small. She pointed to the couch where a blanket was neatly folded—my sleeping quarters for the next week.

"I can't believe you took the week off work . . . for this. You haven't changed one bit," she said.

I laughed. "It's the Nigerian way!"

We jumped onto the foot of her bed and spent the next hour catching up and talking about her career. While I'd struggled to find a promising job in journalism after grad school, she was quickly hired to become a television reporter and weekend anchor at Providence's WPRI. It was a good job in a mid-size market that could easily lead to bigger opportunities, and I wanted to hear every detail.

She told me about the access she had to Providence's underworld as a court and cops reporter. She lit up as she talked about breaking news, working sources, and the pressure of live television. She was especially excited about her time in the anchor chair on weekends.

I wanted what she had. I was grateful to be working in journalism already, but my freelance gig at that news station in the Bronx provided no job security, very little opportunity to grow, and no feeling that I was making a difference in my community. I was ready for a job that mattered, ideally in national news. But I felt woefully unprepared to take that next step.

That's what brought me to Providence that Saturday night in February. I was taking an unpaid week off from my dead-end job in New York to serve as her unofficial intern. It took some persuasion, but Nneka had somehow convinced her bosses to let me shadow her for the entire week. I would be allowed to sit in on morning news meetings where the team pitched story ideas, accompany Nneka and her cameraman in the field, and watch from the studio when she anchored. I had no job leads of my own, no clear path to something better. But I wanted to be completely prepared if and when something came my way. As you know by now, this wasn't something I had come up with on my own. It was the direct result of my mother's devotion to learning in advance.

I spent the coming days traveling with Nneka and her cameraman in the news van wherever they went—the courthouse, the police station, crime scenes, politicians' offices. I grabbed coffee for the team when it was slow and held the light reflector for some of Nneka's live shots. In the evenings, we headed back to the newsroom for the frantic editing and writing process ahead of the nightly broadcast. Nneka and her editor had about an hour and a half to write each story, record the voice-over, and edit the video. I mostly watched, but they sometimes asked me to track time codes from the raw footage to make it easier to find sound bites.

The following Saturday, Nneka took over as the anchor for the 6 p.m. newscast. I stood in the back of the studio trying not to make a sound, taking notes as she navigated the stories about car crashes, court cases, and the state's budget crisis. She asked reporters in the field smart questions and juggled live breaking news. I was entranced by her probing questions and seamless reading of the teleprompter, knowing that producers in the control room were constantly talking in her earpiece.

I gave her a big hug afterward and peppered her with questions. She answered gracefully as she packed up her belongings. It'd been a long day. Still, I couldn't help myself.

"Can I have a go?"

She looked over at the empty set. "You mean anchor?"

"Yeah. To practice. Now that you're off the air."

She shrugged and gestured to the chair.

"Be my guest."

I put my bags down and gingerly climbed into the seat. There I was, in a pair of jeans and a blue hoodie, sitting in a real anchor's chair, facing the same cameras and lights that she had minutes earlier. Nneka walked over to the control room and returned after a few moments, instructing me to put in the earpiece. We obviously weren't on air, but she wanted it to seem real.

"Are you ready?"

Without another word, the crew turned up the lights, played the opening jingle, and cued me in as the prompter began to roll. I cleared my throat and began reading.

"Good evening. Welcome to Eyewitness News at six. We begin with breaking news . . ."

There were various people still milling around the studio—the production crew, a technician, and a sports reporter—some watching out of curiosity and others scrambling to pack up. But for the next fifteen minutes or so, the spotlight was on me. I went through the scripts just as Nneka had. She got the weather reporter to stick around so we could interact "on air." They even had me ad-lib a breaking news scenario with a blank prompter and a producer in my ear to help guide me.

They gave me honest feedback afterward.

"Your voice needs more gravitas."

"Keep your questions short and to the point."

"Don't look down so much."

Variations of the same scene played out over the next two nights. Nneka would come off the air at 6:30, and despite having two more newscasts to prepare for, she and her crew somehow found the generosity to produce a quick fifteen-minute faux newscast just for me. It would have been virtually impossible to create better training conditions on my own. I took it extremely seriously, taking notes after each performance and studying the feedback at length before the next night's show.

Maybe I should have been embarrassed. There I was in my late twenties, my buddy's unofficial intern. I carried equipment, printed scripts, and opened doors. There was no specific job I was training for, no real prospects, and no one in my life actively pushing me in that direction. But there was an old Nigerian saying that floated around my house often when we were little kids: *Life is either pay now or pay later, but if you pay later, there'll be interest.*

I knew that paying now meant practicing being a TV reporter well before I knew of any job openings. It meant staying up late to learn my times tables, hauling buckets of water from the local stream, and two years without television or phone. Somehow I knew all that preparation would groom me for something better. Even if it didn't make sense to anyone else, it would eventually help me. It had to.

Up until that point, the trajectory of my career was underwhelming at best. Aside from freelancing as a TV reporter, I wrote one article a month for a magazine that didn't pay particularly well and was on the verge of shutting down its print publication. Still, *Money* was a well-respected monthly. It featured solid personal finance articles, a talented staff, and a partnership of sorts with CNN through their shared parent company, Time Warner.

I didn't realize the importance of that link at the time. I was much more focused on applying the lessons I'd learned from Nneka to my live shots at my local news gig. I wanted to be better. I started going over my TV appearances and making notes to scrutinize my performance. I printed transcripts from national newscasts I found online and practiced going over them before bed. And I began spending almost all of my Saturday afternoons at the library on Madison and Thirty-Fourth, where I studied everything I could about finance and business for my magazine job.

Nothing happened for several months, but I kept up the routine until one day my magazine boss mentioned something about an opportunity and told me to call him. I dialed his number immediately. It was CNN, he said. The network wanted to hire contributors for a new show. They were looking for someone with a television background and experience at a personal finance or business publication. The show would be on Saturdays, so it wouldn't interfere with my magazine responsibilities. The job description was basically written for me. You could say I was lucky, the right girl in the right place at the right time. But that luck was years in the making.

He asked if I wanted to give it a go. I nodded yes, my eyes wide open. Within days he'd set up a phone call for me with a CNN executive, who invited me to the company's New York headquarters for a screen test and a business news test the following week. The executive apologized for the quick turnaround—he needed to nail down contributors right away, so I'd only have eight days to prepare. He had no idea I had been preparing for an opportunity like this for most of that year.

It may have been the biggest test of my life, but I was relatively calm as I walked into CNN's great glass building by Columbus Circle. I knew I had prepared as well as any human being possibly could have. Still, I started feeling pangs of anxiety as I was led into the main studio on the fifth floor. It was impressive: a sleek glass desk sat under a sea of lights and cameras right in the heart of the sprawling newsroom. There were people everywhere. It was Providence on steroids.

The executive positioned me in front of one of the cameras and handed me a piece of paper with the basics of a business news story about a regional bank merger. There were a few details about the size of the deal, the impact on share price, and a quote from a leading banking analyst.

It didn't say much more, and I was told the test would start in five minutes. That meant I had five minutes to come up with a set of talking points that contextualized the information in a way that would appeal to a national audience: *What would the merger mean for market share? What does this say about broader consolidation within the banking industry?* It was exactly the kind of thing I had been studying at the library and practicing at home.

In a flash, they told me it was time, and the camera's red light flicked on. I looked into the camera and began going through each of my talking points without checking my notes, speaking deliberately in the slower voice I'd been practicing in front of my bathroom mirror since Providence. I desperately tried to ignore the producers who were watching and making notes about my performance. And after only maybe fifteen minutes, it was over.

I obsessively checked my phone and email the next week at work. Patience has never been a strength of mine. It didn't help that my mother called every day seeking updates. I had almost given up when, almost a month later, the CNN executive called my cell.

"I'm so sorry it's taken so long to get back to you. It's been crazy around here."

"No problem!" I was trying to keep cool.

"Unfortunately, we can't offer you the contributor position."

My heart sank. I began to stammer. "Wha-what went wrong?"

"Nothing at all. In fact, we were so impressed by your performance that we want to offer you a full-time correspondent position instead."

I was overwhelmed and slightly confused.

"I'm sorry?"

"We want you to work for us."

When I didn't respond, he chimed in: "You're still interested, right?"

"Definitely!"

"Great. Someone from our team will follow up about the contract."

"Err, yes, of course, umm . . . the contract. Send it over."

I laughed out loud as I hung up the phone. My God. Just like that, only nine months after snagging coffee and holding light reflectors for Nneka, I was going to be a national news correspondent at CNN. I didn't have to call my mother collect this time.

"Mum, guess what?"

"What?"

"I didn't get the contributor position."

"Oh, I'm sorry. That's their loss, darling."

"Wait, Mum. They offered me something better."

"Better?"

"They want me to be a full-time business correspondent. I'm going to work for CNN."

"Oh my." She paused. "What did I tell you, sweets? You can do anything."

My mother passed the phone to my sister, Kandi, who joined in the praise. Also a master preparer, Kandi had just finished her exams in medicine and was on her way to becoming a bona-fide doctor.

"We did it, little sister," I said gleefully.

"I hope Mum lets me watch you on TV," she laughed.

I probably should have been satisfied. I was suddenly a legitimate CNN reporter, doing live shots from the floor of the New York Stock Exchange, interviewing CEOs, and explaining the world's biggest business stories to millions of people around the world. Not to mention that I actually had good health insurance for the first time in my adult life. But I was convinced that if preparing in advance could take me this far, then surely there were no limits.

I had been a business correspondent for barely a year when I reached out to one of the talent coaches at work and asked if I could begin anchor training. The coach, a former professor of mine from Columbia, knew I had been with the company a short time and was confused.

"Why do you want anchor training? Has anyone asked you to anchor?"

"No."

"Then why do you want to be trained?"

"Just in case."

"Just in case what?"

"Just in case an opportunity opens up. I want to be prepared."

He was skeptical, but agreed to help periodically after his official duties were taken care of. We started meeting in a small studio two or three times a week to go through recent shows of the network's prime-time anchors to analyze their performances so I could understand what drew viewers to them.

"Pay attention to how she makes the words her own. She's reading, but she means every word."

"Listen to the range in his voice. There's no monotone. He's drawing you in simply by changing his inflections."

Soon we started reading through scripts on camera, practicing live interviews with stand-in reporters, and navigating breaking news scenarios when the prompter was blank. It meant staying at work for an extra hour or two those days, but as you've learned by now, learning ahead had become something I actually enjoyed.

I could tell that my former professor wasn't so optimistic. He said I was making progress, but he was aware, just as I was, that years can pass without new anchor jobs opening up. And even if one did, it likely wouldn't go to a new correspondent who'd been with the company for barely a year. I kept training anyway, and sure enough, I soon heard from a co-worker that CNN International was looking for new anchors.

Within days I was on the phone with a VP, discussing the position at length. The hours would be brutal. The show was from midnight to 2 a.m. on the overnight wheel, and they were auditioning plenty of other anchors and reporters, with several years more experience than me. She agreed to squeeze me in for a screen test but we both knew my chances were slim.

The test was scheduled for six days later. For a brand-new reporter, one week to prepare for an anchor test would have been a ridiculously short time, but thankfully, I'd been working with my talent coach for several months by then.

He was truly stunned when I shared the news. He always thought it was strange that I was preparing for a job that didn't exist. But now he understood. With less than a week to go, we began practicing every day. And realizing I had gaps in my knowledge of international news, I stayed up late at home studying the countries most likely to be featured on CNN International: China, Russia, India, Saudi Arabia, Cuba, North Korea, and South Africa, among them. I studied their histories, their alliances, and their domestic politics. I was as ready as I could be.

✍

Before I go any further, there's another lesson I need to share. Like most of the lessons in this book, this one was conveyed to me by my mother—and of course, it had been conveyed to her by her parents, and to them by theirs. This one has been critical to my success, and my ability to live at peace with that success.

Put simply, we do not believe in competition. That's not to say it doesn't exist. We are all aware that the world is full of people trying to one-up each other in any and every measure possible. They compete for attention, for money, for status, and for love. I spent much of my childhood competing for self-worth. And as one of the few Black girls at some of the schools I attended, it was a losing battle. The other kids had nicer houses and better clothes. They had two parents. Their hair was silky straight, and mine grew straight up.

Competing with the other kids in those days fed deep insecurities. It also limited my potential. As my mother would say, *it keeps you small.* When I first returned to England after two years in Nigeria, I was painfully aware of the gaps in my knowledge after being part of a different school system. I was initially a C student surrounded by B students. I desperately wanted to be as good as they were, which meant I was still aiming for B's. After grad school, I worked as a $10-an-hour assistant in Los Angeles. I remember complaining to my mother that other assistants made $12 an hour and how much I wanted to be like them.

"You're missing the big picture," she told me. "You aren't competing with them. Prepare as well as you can so you can be your best, not their best."

When I am at my best, I know that I don't need to beat anyone else—or tear anyone down—to reach my highest potential. I practice this imperfectly because I am a human. I feel the pangs of periodic envy like everyone else. But my mother, and my Nigerian roots, help remind me what really matters. I was raised to believe, just as the people of Oghe believe, that a victory for one is a victory for all.

And when I heard there were several talented journalists vying for the same open anchor position, I knew that my success had nothing to do with them. Still, that belief was put to the test when I walked into the newsroom minutes before my screen test was supposed to begin at the company's flagship headquarters in Atlanta. There was a young woman from CNN's New Delhi bureau already there—obviously another candidate for the position. I had seen her on air before. She was brilliant and gorgeous and would have made an excellent anchor.

I introduced myself as I sat down, determined not to see her as a competitor. She confided in me that she was nervous. She worried that she didn't have what it took.

"I know exactly how you feel. But I've seen you on the air and you're incredible. Of course you belong here," I said.

A few minutes later, a young producer tapped me on the shoulder. It was time. She led me through a short corridor into the studio, where there was a large glass anchor desk flanked by three cameras, lights of all different sizes, and a handful of production staff. I remember thinking how warm it was inside the room. I slid into the anchor chair, attached my microphone and earpiece, and shook hands with an executive, who'd run the test and share it with the decision makers.

Minutes later, I was on. The first part of the screen test was simple enough. We covered several stories that I'd prepared for: the ceasefire that had just been announced between Ukraine and Russian-backed separatists, the ongoing battle between Kurdish militia and ISIS, and Hong Kong's umbrella protests.

Then came the hard part.

Twenty seconds into a story about the US midterm elections, the dreaded breaking news test came. A producer in my earpiece informed me that Scotland had just voted to remain part of the UK. The teleprompter went blank. I was expected to ad-lib indefinitely as the control room fed bits and pieces of new information into my ear.

It was a news story that had actually happened a few weeks earlier, so I was somewhat familiar with the details. But it's hard to convey the overwhelming pressure I felt in those moments, knowing that my future at CNN depended on my ability to speak with poise and authority about a complicated topic that was just dropped in my lap.

The words of my talent coach rung in my ears: *Say what you know.*

"Breaking news into CNN. If you're just joining us, the results of the Scottish referendum are in. After a bitter campaign, the people of Scotland have voted against becoming an independent country. First Minister Alex Salmond has called for national unity. We are still awaiting results from a number of key areas . . ."

The Scottish referendum was on the list of topics I'd been studying, so I tried as best as I could to fold in some basic context. It went like that for five minutes—five grueling minutes.

I emerged from the studio mentally and emotionally exhausted, knowing I had done my best but still afraid it wasn't good enough. I had just one more thing to do before I left. The other applicant—my competitor, technically—was still sitting in the same seat, awaiting her turn, a slightly panicked look on her face.

I marched over and told her exactly what happened during my test. I told her about the specific stories they tested me on, the simulated breaking news test on Scotland, what I thought I could have done better. I could tell she wasn't quite sure of what to make of me. Was I messing with her? Was I genuinely helping her?

"We're all in this together," I said. "You'll do great."

My job was done. I walked out of the newsroom feeling light, knowing I had given my best, prepared as well as I could, and genuinely tried to help a colleague. I returned to the hotel room and studied international affairs deep into the night. I suspected, or at least hoped, that there would be more tests. And there were.

They brought me back two days later for another round and a week later to anchor my first live show—the 5 a.m. show on a Saturday—the final test during a low-profile time slot to see if I would sink or swim when it really mattered.

I could think of nothing else in the coming days. Even with the tests complete, I spent every spare moment studying, my notes on global conflicts and world leaders scattered across my apartment. It was a brutal wait. And I didn't get the call for almost a month.

"Thank you for your patience," the voice said when the phone finally rang. "How soon can you move to Atlanta?"

I knew what that meant. The job was mine.

When that brief call ended, I raced to make another one. I was still in shock when she answered.

"From the bottom of my heart, thank you, Mum."

"What? What for?"

"I just got the call. They're making me an anchor. It's really happening."

She screamed, and I screamed, and then we screamed together. Then we started to cry.

"My daughter, a CNN anchor," she said softly. "They will see us now."

Chapter 12

My mother was only five years old when Queen Elizabeth II visited Nigeria for the first time in the 1950s. Despite being a British colony on its way to self-governance, the Nigerian people gave the queen the welcome of her life. My mother had only just started elementary school, but she can still recall the pomp and pageantry that enveloped the country: big cities like Lagos and Kaduna were almost completely shut down in a show of respect, dance troupes from the three main tribes greeted the queen as she stepped onto the tarmac, and young girls lined the roads waving flags as the motorcade slowly rolled past.

Shortly after the queen's arrival, my mother returned home from school one day to find her parents, aunties, and uncles hunched over a small radio balanced against the kitchen window. A woman's voice was speaking, Elizabeth's voice, addressing the nation for the first time as its queen.

Obiajulu had never seen so many adults sitting in silence. As she watched her father hang on to the queen's every word, and her mother careful not to make a sound, she realized for the first time that this place they called the *United Kingdom* must indeed be a kingdom of consequence. That this land, from which this young Elizabeth woman hailed, must be somewhere special.

In the years that followed, she saw how people clamored to send their children to study in Britain, and how Nigerians who returned from London walked with a certain swagger. There were clearly opportunities in the West that didn't exist in Enugu or Jos. Eventually, she began to dream of a life for herself there, too.

The trauma of the Biafra war rattled her ambitions for a time. There was so much to do to rebuild her homeland after the war; schools needed teachers, hospitals needed nurses, and the community needed support.

Up against the need to help her people, moving to England seemed as plausible as moving to Neverland. But eventually, and after much persuasion from my father, she agreed to leave—only after promising to return someday.

The majesty of London was intoxicating at first. In the weeks after they arrived, my parents posed for pictures outside Big Ben, the Tower of London, and Buckingham Palace's grand gates, debating how many rooms there were and how they were decorated.

But amid all the grandeur, she felt lost inside. She desperately missed the sense of community that had enveloped her for as long as she could remember back home. That was one of the great struggles of that time. No matter how hard she tried to connect with other people in London, it just didn't happen. Outside our South London home, she almost never felt that she belonged, that she was welcome. I never heard her complain, but I know it bothered her.

And so, when the queen invited my mother to Buckingham Palace in 2008, I can't begin to explain to you how much it meant.

She almost collapsed in disbelief when Chiwetel handed her the invitation. She staggered onto the couch, fingering the bold black lettering with the Lord Chamberlain's signature to make sure it was real.

She knew that between the lines of that single sheet of paper was a powerful message: the people of Britain had finally seen her. After years of being dismissed and overlooked, her adopted countrymen had come to appreciate what she—and by extension, the people of Oghe—represented.

Although she'd lived in Britain for decades by then, Obiajulu knew very little about what an Order of the British Empire actually was. She'd seen the letters *OBE* and *CBE* alongside famous names in newspapers, and they looked important, but she'd read right past them for years.

When she learned Chiwetel had been invited to join this exclusive club, and that she'd be right there at Buckingham Palace to witness the ceremony, she made it her mission to learn as much as she could. Over the course of her intense research, she was proud to learn that the queen bestowed such honors on those who'd made outstanding contributions to the arts, sciences, and culture, a group that included Paul McCartney, that same Beatles rock legend Arinze had idolized when he was a schoolboy in Enugu.

In the weeks leading up to the ceremony, my mother kept the invitation within arm's reach at all times. She would sometimes stop in the middle of frying plantains or scrubbing the sink to run her fingers over the thick paper and its bold lettering. More than once, she abruptly ended a phone conversation just to make sure it was still there.

She awoke the day of the ceremony afraid it was a dream. She was, after all, a village girl deep down, and village girls were not invited to the queen's palace. Yet here she was, invitation in hand.

She thought about Arinze as she dressed that morning, wishing he could see how far they had come. It was really only because of him, his dreams and ambition, that Chiwetel had developed a passion for reading prose and poetry in the first place, and because of him that her children were well versed in great playwrights like Stoppard, Pinter, and Shakespeare. To thank him, and her village, she decided to wear a traditional African wrapper and blouse with a glimmering gold ichafu around her head. She'd rummaged through a market stall in North London to find something appropriate, something to show the queen and the world that she was unmistakably Nigerian. After years of trying to fit into British society, this time she wanted nothing more than to stand out.

That morning, a black Mercedes pulled up to the same Norwood home where we were raised. My mother walked slowly to the car as if to savor every moment of her fairy tale. We piled into the back, with her in the middle, surrounded by her brood of smiling grown children. She let out a deep sigh, and we were on our way.

We motored through the streets of South London up through Brixton, close to the pharmacy she'd recently sold, and then past Stockwell, where she and my dad once shared a rundown one-room flat. We crossed the River Thames into Victoria before stopping at Buckingham Palace's front gates, a black-and-gold symbol of strength for one of the most powerful kingdoms the world has ever known.

My mother had never seen the gates open. None of us had. They were closed when she and my father gazed upon them during that first sightseeing mission in the early '70s. They were closed every time she passed on her way to the University of London's pharmacy school. And they were closed still whenever she drove by on her way to Shepherd's Bush Market to buy kids clothes in those stores where her money went a little further.

We showed the palace guards our invitations, and slowly but surely, the gates of Buckingham Palace opened to my mother.

🖎

Some people's life stories are written on their faces. Tucked within the lines, undulations, and curves sits a tragedy or triumph, a lost love or a resounding joy. For as long as I could remember, my mother's face told a story of pain and grief. Even when she'd smile, I could see years of war, loss, and fight peeking out. But that morning, as we drove onto the palace grounds, we detected something different in the lines around her eyes as she stared out the window. It's probably too simple to call it happiness or pride or peace. Whatever it was, my mother's face seemed lighter as she stepped onto the forecourt of Buckingham Palace and adjusted her Nigerian head wrap.

Led by Chiwetel, we followed a small line of VIPs past the smartly dressed staff and into the building. We moved through a vast hallway adorned with oil paintings, plush wallpaper, and an army of butlers that flowed into a palatial auditorium already half filled with the other honorees' families. My mother disregarded any nerves and marched directly to the front row, where she claimed a seat in the middle. She had earned that moment, and that seat, and she wasn't going to miss a single thing. We didn't realize until afterward that her giant Nigerian head wrap was blocking some of the other guests' view.

My mother has never been a particularly nostalgic person, but the details of that day are forever imprinted on her mind. She actually wrote them down later in the day lest her memory one day betray her. It hasn't yet. She can recall in vivid detail the moment Prince Charles entered the room in his decorated military uniform. Minutes later, her son was standing face-to-face with the heir apparent to the British throne. They spoke softly for a few moments before Charles handed Chiwetel a small golden medal with a red ribbon and helped pin it to his lapel. My mother was too emotional to join the polite applause that filled the room, sitting perfectly still as tears fell.

When it was over, we walked together past the marble statues, the gold-framed portraits, and the bronze candelabras. My mother linked arms with Chiwetel just as she dabbed her moist eyes with a tissue, and they shared a soft smile. In her other hand, she clutched his medal, a golden cross gifted by the matriarch of Great Britain herself. The queen could not have known how truly unlikely that moment was—how my mother had barely survived war all those years earlier, how Chiwetel had been left for dead in the back of a truck, how she'd nearly succumbed to grief after losing my dad, and how, as a poor immigrant in a society that barely saw her, she'd scraped and clawed and carried her children on her back to give them a better life.

The black Mercedes was waiting for us in the forecourt after the ceremony, and as we approached, the driver opened the door. My mother paused, glanced back at the palace, the queen's residence itself, and turned to face us.

"You never know where your children will take you."

Epilogue

My grandmother, known to the world as Caroline Usonwa Okafor, passed away on August 26, 2019, at home in Enugu. Her birthday wasn't recorded when she was born, but we think she was eighty-one, maybe slightly older.

Mama Nnukwu had lived a long life, a good life, but I sobbed for hours all the same after learning she'd left us. I so wanted my children to meet the woman I was shipped back to as a little girl, who taught me the ways of the village and taught my mother to be the parenting force that carried us after Dad died.

My siblings and I flew to Enugu for the funeral, the first my mother had planned since Dad's. We packed together in Mama Nnukwu's modest living room the night before the service, crying, laughing, and sharing stories about how she had shaped each of us in some special way.

I joked about how she'd watch me like a hawk at dinner to ensure I ate every morsel of the Nigerian food she'd prepared. If I didn't, I'd go to bed hungry. Obinze told us how she'd refuse to speak to him unless he spoke Igbo. And my mother recalled through tears how Mama Nnukwu had slept on the muddy floor of the Enugu train station waiting for Grandpa to escape the pogroms. She refused to leave, even when her children begged her to.

My grandmother had served as her family's fiercest protector, just as my mother had, more drill sergeant than nurturer, but someone who had quite literally given everything she had to family and community. She would go days without eating during the war to make sure her children had what they needed. And later, she welcomed me into her home for two years, shared her crowded bed, her discipline, her love.

We woke up at dawn on the day of the funeral, dressed all in white, as is the custom, and headed to the morgue. As we parked the car, I noticed a look on my mother's face I hadn't seen in a long time—a look of deep dread.

"The last time I was here"—she paused—"was for your father."

She continued: "It's the same place they took your brother when they thought he was gone, too. I don't like it here."

Very little had changed since her last visit, she told me. There were the same shanty stores across the street, the same stream of hawkers selling bananas and groundnuts, and the same dilapidated security shack that guarded the parking area. She held her breath for a moment, shaking with emotion, desperate not to cry anymore. I reached for her hand as we stepped out of the car toward the morgue, shoulder to shoulder, waiting for the men to load the casket into the hearse.

Looking over at my grieving mother on that dusty road, I realized just how much our lives had changed. The last time she walked this path, she was a hopeless, pregnant widow, the mother of three young children consumed by grief and unsure how to survive in a foreign land.

Almost three decades later, the boy once left for dead in that morgue had become an Academy Award nominee; the baby she carried in her womb, Kandibe, had never met her father, but followed in his footsteps by becoming a doctor in London. The daughter standing beside her was now a global news anchor. And Obinze, then a troubled teenager, had landed on his feet as a successful businessman.

We rode together in silence to my grandma's childhood home before moving to the church for the formal service, where we paid our respects at the open casket.

My mother stood and addressed the crowded congregation after the priest delivered the opening prayer. She spoke of her mother's sacrifices, her fiery commitment to family, and the extraordinary steps she took to help her children overcome the horrors of war. She spoke of the school lessons in the bush, the bamboo shelters, the late-night lullabies to drown out the sound of gunfire.

"Mama remained a gentle but formidable character," my mother said, her voice trembling. "And she was always a great ally in your corner to help you fight."

That last word stuck with me for the rest of the service as I sat next to my brother in the wooden pew. *Fight.* Of course, Grandma was a fighter. We are nothing if not fighters. We fight for what we want from life, we fight the forces that stand in our way, and most of all, we fight for our families.

After the final prayer, my mother and I walked arm in arm out of the church and back toward Mama Nnukwu's house, the same house where my teenage father had begged for her blessing a lifetime ago. I squeezed my mother's hand as her mother's coffin was lowered into the empty plot in front of her home. The look of terror in her eyes from earlier in the day had been replaced by a sad acceptance. She knew, as I did, that Mama Nnukwu had left her mark on the world. She, too, had made her village proud.

The next day, as friends and extended family finished cleaning the funeral site, and the uncles and aunties packed away their white shirts and dresses, my mother and I took a drive through Enugu. We passed by the slow-moving waters of the Mmili Ocha, the colorful stalls of the Abakpa market, and the barren campus of my elementary school. Eventually, we pulled into a dirt parking area outside a modern two-story building in Trans-Ekulu.

It was the same plot of land where my father once told my mother he'd like to build a school. It wouldn't be until after he had found success in London, of course, but when the time was right, Arinze would return home to ensure the children of his community received the kind of education he and my mother were robbed of during the war.

My father's life ended long before he could realize such dreams. But my mother never forgot them. And a few years before my grandmother's death, my siblings and I chipped in to help make them a reality.

Today, Obiajulu Justina Ejiofor, a woman who finished high school starving in the mud, is the founder of Brightland, a school that hands out scholarships to children fleeing violence from Nigeria's troubled northern region—just as she once had. The school boasts an athletic field, computer access for each child, and a modern science lab under construction. Such amenities may be common where you live, but they are truly transformative in Enugu.

We walked through the front gate that day, where the security guard greeted us with a reverential nod, and moved across the newly installed sports field to a tented area where a few dozen ten- and eleven-year-olds were sitting quietly in rows of chairs. We shuffled toward them slowly, weighed down by heavy hearts and puffy eyes. My mother had asked me to speak to the students that day, and as raw as I felt, I was happy to share my story.

It was easy to see, as we walked to the small platform at the head of the group, how far we had come. The school's very existence was a testament to my family's successes and the strong women—Mama Nnukwu at the top of the list—whose village ways made them possible.

My mother and I turned to face the children, a new generation of my countrymen, who were silent and smiling as my mother began to speak.

"This is my daughter," she told them. "Listen to what she has to say."

Acknowledgments

I cannot begin to put into words the gratitude I feel for my parents, Arinze and Obiajulu Ejiofor. Thank you to my incredibly supportive siblings, Obinze, Chiwetel, and Kandi, for encouraging me to share Mom's story with the world. To my husband, Steve; this book would have never happened without your incredible support and editing skills. To my beloved grandparents Caroline and Ignatius Okafor, thank you for bringing me into your home and showing me the beauty of Nigerian culture up close. Thank you to my agent extraordinaire, Kirby Kim, who encouraged me to write this story (I'm so glad I listened), and Clare Conrad and Eloy Bleifuss.

I also had a team of friends and family who offered critical feedback throughout the writing process: Cinque Henderson, who held my hand from the proposal all the way to the galleys, Cin Fabre, Cynthia Peoples, Thomas Franklin Lane, Ayesha Harruna Attah, Kate Thomas, Mohsin Drabu, Nnanna Ude, and Lawrence Ijebor (my dad would be so happy our paths crossed). Many others walked me through Nigerian

Acknowledgments

history, culture, and shared their fondest memories of my dad: Edwin Okafor, Leo Okafor, Tony Tagbo, Tony Obuaya, Mike Obiekwe, Sylvester Amamilo, Antonia Obiekwe, Chioma Igboeli, Okezie Igboeli, and Mary Igboeli.

To my editor, Tracy Sherrod, thank you for believing in this story. My amazing team at HarperCollins US: Judith Curr, Tara Parsons, Brieana Garcia, Maya Lewis, Courtney Nobile, Sarah Schoof, Jennifer Baker, Andy LeCount, Yvonne Chan, Stephen Brayda, Mary Grangeia, and Alexa Allen. Thank you to my HarperCollins UK family: Michelle Kane, Helen Garnons-Williams, Jo Thompson, Katy Archer, Nicola Webb, and Lindsay Terrell.

To all my uplifters, thank you for changing my life: Femi Oke, John Chinegwundoh, and Nneka Nwosu. Thank you to my supportive family at CNN, including Jeff Zucker, Mike McCarthy, Meara Erdozain, Penny Manis, Lauren Cone, and David Brandt (who read my proposal back in 2019 and said it had potential). I'd also like to thank Beowulf Sheehan, Yoko Fumoto, Regina Chinegwundoh, Frank Chinegwundoh, Veronika Kovacs at Keble College, Oxford, Barb Hall at TH Entertainment, Tope Esan, Afam Onyema, Kingsley Chiakwa, Kelly Burns, Elias Schulze, and Ike Anya.

Finally, a huge thank you to Noah and Louis for being on your best behavior while Mommy wrote this.